Understanding Sun Tzu
And
The Art of Hybrid War

PETER LANG PROMPT

Zia Ul Haque Shamsi

Understanding Sun Tzu
And
The Art of Hybrid War

PETER LANG
Lausanne • Berlin • Bruxelles • Chennai • New York • Oxford

Library of Congress Cataloging-in-Publication Control Number: 2023026865

Bibliographic information published by the **Deutsche Nationalbibliothek**.
The German National Library lists this publication in the German
National Bibliography; detailed bibliographic data is available
on the Internet at http://dnb.d-nb.de.

Cover design by Peter Lang Group AG

ISBN 9781636672335 (hardback)
ISBN 9781636672632 (ebook)
ISBN 9781636672649 (epub)
DOI 10.3726/ b21042

© 2023 Peter Lang Group AG, Lausanne
Published by Peter Lang Publishing Inc., New York, USA
info@peterlang.com—www.peterlang.com

All rights reserved.
All parts of this publication are protected by copyright.
Any utilization outside the strict limits of the copyright law, without the permission of the
publisher, is forbidden and liable to prosecution.
This applies in particular to reproductions, translations, microfilming, and storage and
processing in electronic retrieval systems.

This publication has been peer reviewed.

Dedicated to
Soldiers and Civilians
Who lose their lives during Unnecessary Wars?

The supreme art of war is to subdue the enemy without fighting.

Sun Tzu

Win wars by other means

Zia Ul Haque Shamsi

CONTENTS

Preface	xi
Acknowledgment	xiii
Introduction	1
Chapter 1 The Art and Science of Hybrid War	13
Chapter 2 Avoid Prolonged Wars	23
Chapter 3 Know Your Enemy and Know Yourself	33
Chapter 4 Significance of Civil-Military Relations	43
Chapter 5 Defeating the Victory	53
Chapter 6 Significance of Intelligence Operations	63
Conclusion	73

Appendix A	79
Appendix B	89
Bibliography	93

PREFACE

The concept of hybrid warfare is as old as the warfare itself. However, it has re-emerged with an expanded scope in the twenty-first century. The purpose of the entire campaign is to break the will of the people and win the war without fighting. The modus-oprandi of the planning and execution of today's hybrid warfare, particularly against unequal military powers (UMPs), sounds much similar to the precepts of the Chinese sage, Sun Tzu, who prophesied of winning the war without fighting. An analysis of Sun Tzu's *The Art of War* reveals that he was a proponent of expanding state's influence without the use of force. Inspired by Sun Tzu's precepts, this author is proposing that states may acquire power for the promotion of peace within and peace without, as opposed to classical realism which insists on power and security. The sustained period of peace would enable the state to concentrate on economic development and societal progress. The same would help the state in perception management and project its soft power among the comity of nations. Such a state would then have the capacity to help develop other regional countries to expand its area of influence. However, the states that use their hard power to subdue other nations not only violate Sun Tzu's dicta; end up spending huge resources, cause total destruction of the target state, and yet unable to maintain a sustained occupation in contemporary international system.

In view of the continued relevance, this author attempts to rediscover Sun Tzu's precepts from *The Art of War*, and its application in today's hybrid warfare, deploying deductive reasoning, and qualitative analysis of contemporary wars and conflicts.

ACKNOWLEDGMENT

By the Grace of Allah, my third international book, fourth otherwise, and that too on my favorite subject is available now. I bow my head Before Allah Almighty for His Countless Blessings upon me during the entire period of this project.

I am indebted to all academics, and practitioners from military, diplomatic and civilian bureaucracy, whose talks, lectures and discussions have benefited me immensely in shaping my research on this extremely important subject. An effort has been made to acknowledge and refer their work appropriately; however, if any particular idea or work is not referred properly, I would seek guidance to make suitable correction. I must also highlight that some of the ideas presented in this book have been published in *Daily Times*, Pakistan, and *The Nation*, while this project was underway. However, I have tried to refer them where required.

I am also grateful to Peter Lang Publishing, Inc. for the continued support in completion of this project. The idea is to generate discussions, deliberations and awareness about the dangerous consequences of the employment of hybrid warfare in contemporary wars against UMPs by the relatively powerful states. And, in the process, if we can stop the next war, I would have achieved my objective.

Thanks again to my family for supporting me in my academic pursuits.

INTRODUCTION

Hybrid warfare, as a concept and military strategy, is as old as the warfare itself. Though the term *hybrid warfare* was first used in 2002 by Major William J. Nemeth, when he examined Chechen Insurgency.[1] Chifu and Anghel have quoted Nemeth's thesis, "the 'hybrid' fusion between religion, society and modernization, namely technology in the hybrid society results in the postulation of hybrid war/conflict."[2] The concept employs a combination of political warfare alongside conventional warfare, irregular warfare and cyber warfare. It also deploys variety of methods to instigate the people of the target state through media campaigns propagating fake news, diplomatic maneuvers, manipulation of the workings of international institutions, and even interfering in its electoral system. The purpose is to alienate the people from state institutions and weaken them from within.

While there is little debate that there is nothing novel about the concept of hybrid warfare, however, there is agreement that it has re-emerged with an expanded scope in the twenty-first century. Hybrid warfare today seamlessly combines a number of elements: kinetic and non-kinetic, at any given time against the target state. In fact, the kinetic application is deployed as the last resort, while non-kinetic elements that include economic warfare, psychological warfare, media campaign, etc., act as the frontline assault by the perpetrators, which may be state versus state, or the state versus non-state actors, or a combination of the two.

The purpose of the entire campaign is to break the will of the people and win the war without fighting by projecting the incumbent regime as toothless, incompetent, and rouge, defying the international norms, practices, and universal democratic values. The modus-oprandi of the planning and execution of today's hybrid warfare, particularly against UMPs sounds much similar to age-old Chinese sage, Sun Tzu's precepts, who some 2,500 years ago, insisted on winning the war without fighting.

This book aims to carry out an in-depth analysis of Sun Tzu's famous treatise *The Art of War*, and apply his dicta in today's hybrid warfare by global and regional powers against relatively weaker military and economic powers usually referred as the UMPs. This author attempts to rediscover Sun Tzu's precepts and its application in today's hybrid warfare, deploying deductive reasoning, and analyzing the practical outcomes of the selected events.

Sun Tzu and his *The Art of War* are regarded as an important document on state, policy, strategy, and tactics in war. The precepts of the Chinese sage are widely quoted by politico-military leadership in their writings as well as presentations. However, it is intriguing to note that stakeholders around the globe do not really follow his dicta when it comes to avoidance of wars and winning without fighting. Hence, the analyses of most wars and conflicts reveal that the objectives could be achieved without violence, simply following Sun Tzu's precepts. Whereas, any such event can be studied for the purpose to prove this author's assertions, this research would focus on his dicta of winning wars without fighting, and peaceful expansion of state's power employing relevant tenets of hybrid warfare.

Evidently, the hybridity is inbuilt in the application of warfare, particularly when Sun Tzu's precepts are studied. His insistence on making an all-out effort to win the war without fighting, and avoidance of prolonged military campaigns, clearly reflects the use of non-kinetic approaches at first and the use of force as a last resort, and that too for a very short time.

Hoffman's definition explains the concept as follows:

> Hybrid Wars incorporate a range of different modes of warfare, including conventional capabilities, irregular tactics and formations, terrorist acts including indiscriminate violence and coercion, and criminal disorder. These multi-modal activities can be conducted by separate units, or even by the same unit, but are generally operationally and tactically directed and coordinated within the main battlespace to achieve synergistic effects.[3]

While the core concept of hybridity in warfare may not be as contemporary, but its expanded scope makes it more unique in the changed paradigm. Hybrid warfare is generally executed through four common elements which cover the expanded spectrum of present-day warfare: conventional warfare; irregular warfare (i.e., counterinsurgency and terrorism); related asymmetric warfare (i.e., unconventional warfare as partial warfare); and composite warfare, whereby irregular forces are deployed concurrently against an adversary and used by state actors to complement normal warfare approaches.[4]

In fact,

> hybrid warfare became a buzzword in the international political discourse following Russia's invasion in Ukraine and the illegal annexation of Crimea in 2014 ... on the Kremlin's use of hybrid tactics because the Kremlin's hybrid toolbox is arguably the most sophisticated and comprehensive. Russia's hybrid warfare also primarily targets the Euro-Atlantic community and the countries in the "grey zone" between NATO/EU and Russia.[5]

The perpetrators of hybrid war aim to create a synergetic effect of their effort on the target state, utilizing all available avenues: tangible and intangible. Tangible elements of warfare which are far more scientific now due to the technological revolution in military warfare, and sophistication in weapon inventories, include physical impact through military hardware where the enemy is visible. However, the intangible elements of warfare which fall in the domain of artistic applications, and where the enemy is invisible and operates from within, is far more damaging and effective. This domain of hybrid warfare has an exceptionally large canvass: for instance, economic warfare may include disruption in stock exchanges and creating volatility in the currency market, both aimed at purposely scaring investors away to weaken a state.

The hybrid warfare is launched in synergy with media campaign, economic coercion, cyberattacks, espionage, political interference, electoral engineering, and limited military intervention, if required. In fact, "since the armed confrontation between Hezbollah and Israel in 2006, the concepts such as 'hybrid war,' 'hybrid threat' and 'hybrid adversary' have been on the rise. These terms effectively ... combine conventional and non-conventional fighting capabilities, and who possess quasi-state characteristics."[6]

This author has introduced the phrase "dissidence, despondency, and disappointment (D3)" among the populace of the target state, as part of the hybrid campaign.[7] By doing so, the perpetrators will be able to cause great harm to the target state without raining alarm bells among the international institutions, and with considerably reduced violence and perhaps the expenditures on the campaign efforts. The spirit behind D3 on part of the perpetrators would be to create unrest in the country, and doubts in the minds of the people that the state is failing to protect their interests, lives and properties. In most cases, it would be done through the local operators as was done by India against Pakistan in the last two decades, and revealed by European Watchdog DisInfoLab through Indian Chronicles.[8] As many as 750 fake websites were created to defame Pakistan and China in the western capitals. By doing so, India was able to avoid a worldwide criticism and could easily reject Pakistani allegations on its activities inside Pakistan, particularly in Karachi and Balochistan. Likewise on the economic front, it has now been accepted by Indian officials that they were resisting for the removal of Pakistan's name from the gray list of Financial Action Task Force (FATF) even after the country had complied with majority of the action points required by the watchdog.

The debate on the origin and the definition on hybrid war continues: "Hybrid War is badly defined and it is often unclear what is being argued. Hybridists insist they have identified a new kind of war. There is nothing new here. ... Means and methods, largely tactical, describe part of the nature of the war, but they do not give us a new form of war."[9] However, it is worth noting that fierce debate in the literature surrounds the concept of hybrid conflicts. Many experts maintain that warfare has always assumed a hybrid character throughout the history of international conflict—Sun Tzu's way. However, most specialists assert that hybrid warfare is a relatively new phenomenon in world politics. Increasingly, nations face new forms of interstate confrontation and effective, powerful tools for strategic non-nuclear deterrence that constitute a complex of power measures and impact on rival nations.[10]

This academic debate on the definition of hybrid threats enables researchers to grasp the term's multifaceted nature, while also presenting examples of hybrid threats such as terrorism and migration. The same report underlines that "hybrid threats are not exclusively a tool of asymmetric or non-state actors, but can be applied by state and non-state actors alike. Their principal attraction from the point of view of a state actor is that they can be largely non-attributable, and therefore applied in situations where more overt action is ruled out for any number of reasons."[11]

Though hybridity in threats against an adversary is not a new strategy in non-kinetic warfare, it has certainly paralyzed the international law in its application to avert the threat, particularly against relatively smaller and weaker states. Perhaps, this is one reason that Europe is faced with another war at the moment, which clearly reflects the inability of international laws to avert kinetic and non-kinetic threats of different nature and character. However, Baytuk is of the view, that most probable driver of the emergence of hybrid threats is grounded in the evolving nature of global order, of which the social, political and economic components have changed dramatically since the end of the Cold War.[12] Perhaps, Andrew Mumford's description of a hybrid adversary is most comprehensive. According to him, a hybrid adversary is "one that uses a combination of political, military, economic, social and intelligence methods of influence, as well as conventional, irregular, terrorist and criminal methods of warfare."[13]

As hybrid threats to international security have evolved, their use in scholarly and policy debates has become a source of ongoing confusion. In many instances, it can be noticed that concepts such as "hybrid threat" and "hybrid war" are used randomly, without even a working definition provided for the terms. This has led to further confusion for policymakers instead of much-needed conceptual clarification.[14]

The international system is fast evolving with new political alignments. While the United States (US) wants to extend its superpower status through the exploitation of hard power, China is striving to achieve the same through the elements of soft power. A clear demonstration of this author's assertions was seen on March 10, 2023, when China hosted and facilitated talks between Saudi Arabia and Iran, which resulted in two Middle Eastern arch rivals in agreeing to resume diplomatic ties and open their Embassies in Riyadh and Tehran within two months. It is necessary to mention that "Riyadh cut ties with Tehran after Iranian protesters attacked Saudi diplomatic missions in Iran in 2016 following the Saudi execution of the revered Shia cleric Nimr al-Nimr. The rivalry between predominantly Shia Iran and Sunni Saudi Arabia has dominated Middle East politics in recent years, spreading into Syria, Iraq, Lebanon and Yemen."[15] Apparently, China is depending on the art of war, whereas the US is deploying the scientific elements of modern-day warfare. The question remains as to where do Europe and Russia stand at the moment? Perhaps, the answer would come when the iron curtain is finally drawn between Russia and Ukraine, unfortunate though, but highly probable at the moment.

When this author first defined WAR as "Waste of Available Resources,"[16] few people reacted to it by calling me an idealist or liberalist or even constructivist. However, most of them did understand what I meant. It is understable that warfare remains the most ancient form to resolve disputes, whereas rest everything has been transformed. Yet, Bilateral Treaties, Agreements, Multilateral Treaties, League of Nations, and United Nations have not been able to avoid or prevent wars between states with territorial disputes. However, most painful military engagements are between UMPs, even if they are not for the territorial disputes, particularly in modern times.

The end of the Cold War deprived the sole superpower of the time, the US an enemy, but Iraq's President Saddam Hussain readily provided a battleground in the Middle East by invading Kuwait, and inviting the US and its allies to begin a new era of wars and conflicts between UMPs. Iraq was destroyed even if Saddam survived after the First Gulf War, only to be demolished again a decade later.[17] Likewise, Osama Bin Laden's presence in Afghanistan at the time of 9/11, provided enough justification to destroy an already war-ravaged state completely and its people to ruins forever, perhaps.

It may be necessary to mention the famous dictum of erstwhile Soviet leader Andre Gromyko: "Ten years of talk is better than one day of war."[18] Gromyko was the longest-serving Foreign Minister of the Union of Soviet Socialist Republics (USSR) during the Cold War, and his precepts about war and its terrible consequences support this author's definition of WAR as the "Waste of Available Resources." Because, each major war is usually followed by some Treaty, Agreement, or a Marshall Plan to rebuild the destroyed state. However, giving some weight to realist paradigm, a healthy competition between rival states for the sake of development in science and technology, education and health services to their people, culture and sports, tourism, and interconnectedness would be encouraged, but not WAR, at any cost or reason.

Perhaps, in the same vein, the 2011 Jasmine Revolution in Tunisia brought in certain changes in the thought process of the people of the region. The people showed the desire for democratic governments instead of living with dynastic rules. However, it succeeded in few countries only. While Bahrain could endure the uprising and resettled with old-styled governance, and Syria remains at an internal war ever since. Libya has changed for good perhaps, but Egypt returned to age-old military-styled government after a brief period of Morsi's rule. These events were largely supported by the proliferation of terrorist groups in the region. Since the use of kinetic and cyber threats as responses to global terrorism has blurred the traditional differences between peace and war, these types of conflicts have been categorized under the umbrella of hybrid warfare.

Elsewhere in the Middle East, The Islamic State of Iraq and the Levant (ISIL) is one non-state actor that has used hybrid tactics against the Iraqi military. Moreover, ISIL has retained its provisional ambitions and has engaged in regular and irregular acts of terrorism. In response, Iraq has transformed its strategy to favor using hybrid tactics against international and non-state actors. Moreover, the US has also used a range of conventional airpower and other tactics on Iraqi government troops. Hybrid war is a conflict between state groups and non-state actors who pursue overlapping objectives and weak domestic states.

In the European theater, it is alleged that Russia has used refugees to infiltrate Europe in order to undermine regional economies and foment social turbulence. Finland Parliament has amended legislation on border security that "allows closure of crossing points with Russia amid fears that Moscow could choose to send large number of migrants to the frontier .... The risk of such hybrid threats from Moscow is seen as being particularly high now that Finland has become an observer member of NATO...."[19]

The entire debate brings back the original argument as dictated by Sun Tzu that the acme of skill lies in winning the war without fighting. This comprehensive definition of hybrid threats enables researchers to grasp the term and proceed on a wide-ranging canvass of warfare that remains the most expensive method of resolving the disputes.

Central Argument

The concept of hybrid war is as old as the warfare itself, though the terminology may be relatively newer. The spirit of hybrid warfare lies in the Chinese sage, Sun Tzu's dictum of winning the war without fighting. Also, the adoption of hybrid applications in wars and conflicts is heavily inspired by his precepts on statecraft defined in his well-read and widely referred treatises: *The Art of War*. Therefore, this book is aimed at carrying out an in-depth analysis of Sun Tzu's dicta through the lens of hybrid mannerism of warfare in the contemporary environment.

The significance of this research lies in its purpose in reiterating the need for peace, stability, and security through the avoidance of wars, in the changed paradigm, primarily because the number of wars and conflicts between UMPs has been on the rise since the beginning of the twenty-first century.

The other objective of the research is to fill the gaps in the scholarship of this extremely important subject of hybrid applications during peace and war. This author maintains that WAR is "Waste of Available Resources," perhaps the idea is closely related to Sun Tzu's precepts of avoiding wars and winning without fighting. Therefore, the security is essentially needed to safeguard the resources for the people of the state, and not wasted on the needless rivalries. This may sound idealistic, but it is more pragmatic and certainly doable to save the humanity from the impending Armageddon.

Intellectual Puzzle

Another intellectual puzzle that this research aims to investigate is the strategic dichotomy between peace and security. Should states acquire power to ensure security through the use of force in wars and conflicts, or strive for peace within and peace without by avoiding wars and conflicts with utmost preparation for the worst? Understandably, this would increase reliance on unique deterrent capability of the state through strong military and robust economy. Perhaps, Sun Tzu meant the same.

Theoretical Framework

Realism would form the basis of the argument of this research to determine warring states' behavior. However, the detailed analyses of Sun Tzu's precepts through the lens of hybrid applications and approaches in contemporary wars and conflicts would be undertaken.

Methodology

Based on the deductive approach, and following the principles laid for the research on subjects related to social sciences, this research undertook qualitative analysis making use of primary as well as secondary data.

Plan of the Book

Sun Tzu's *The Art of War* is a concise but a comprehensive document, comprised of thirteen chapters. However, the volume is not very thick, and its translations in over 100 languages are easily comprehensible. Each chapter selects a particular subject and briefly outlines the guidelines on policy, strategy, and tactics: both for the state and for the armed forces.

INTRODUCTION

This book aims to carry out an in-depth analysis of Sun Tzu's famous treatise *The Art of War*, and apply his dicta in today's hybrid warfare by global and regional powers against relatively weaker military and economic powers, or the UMPs.

Introduction contains the glimpse of a comprehensive debate on the concepts of the hybrid warfare. It also presents comparative definitions of the contested subject. The central idea, the intellectual puzzle along with the significance and objective of the book are also outlined to understand the purpose and the concept.

Chapter 1 of this book, "The Art and Science of Hybrid War," is aimed at building up the argument that augments the efficacy of hybrid war and its employment based on art and science. It highlights the reasons why hybrid application of warfare is becoming so popular in contemporary international order, even though the concept is as old as the warfare itself and dates back to Sun Tzu's era.

Chapter 2, "Avoid Prolonged Wars," explores the cardinal principles outlined in chapters I and II of the Sun Tzu's book. The most important aspects of chapter I from his book are: Secrecy in Planning, Pretention, Surprise, Utmost Preparation, and Comparative Analysis. Whereas, chapter II of Sun Tzu's book dilates on an all-important factor of his insistence on avoidance of a prolonged war. The other important aspects that this author intends to analyze include: War is dangerous, and expensive with high inflation, for which the resources would always be inadequate. However, if war is unavoidable, a short, swift, and quick decision is strongly recommended. Sun Tzu insisted that one should aim only for victory, and motivate the troops against enemy and reward them after victory. He also laid great emphasis on treating prisoners of war (POW), humanely.

Chapter 3 of this book, "Know Your Enemy and Know Yourself," would analyze chapters III, IV, and V of *The Art of War*. These chapters are significant due to the contents related to war itself. Sun Tzu had long insisted on wining without fighting and if necessary a short, and fierce attack for a quick victory. In the domain of hybrid applications, he insisted on striving for the psychological ascendency, and avoid killing the enemy soldiers, rather insisted on seeking surrender and include them in your force. Sun Tzu laid great emphasis on preparation and surprise, expanding the capacity to fight, and above all the non-interference from political leadership. Sun Tzu's famous dictum that is widely quoted, "know your enemy and know yourself," would also be analyzed through hybrid applications in contemporary wars and conflicts. Since Sun Tzu was against prolonged wars, he vehemently opposed the siege and declared it as a bad strategy. In the following chapters, Sun Tzu insisted on the soundness of the plan that is offensive, multidirectional, and hence unpredictable.

Chapter 4 of this book deals with chapter VI, VII, and VIII of *The Art of War*. Sun Tzu demanded creativity in strategy by the commanders. He was proponent of the delegation of power and authority, and formation of different commands. Interestingly, he prophesied pretention, perhaps opposite to what deterrence demands; strong show of intent. However, he insisted on keeping enemy engaged and continue relentlessly planning of attack secretly.

The fifth chapter of the book analyzes the contents of chapter IX and X from Sun Tzu's book. He talks of the policy matters between various organs of the state. He warned that if the state organs are not on the same page, forget about war, and if the armed forces are not one, you cannot win a war. Sun Tzu laid the foundations of clear delineation between policy and strategy. All policy decisions must come from the ruler with allocation of resources, and the strategy must be formulated by the commander to accomplish the defined ends. In the following chapter, Sun Tzu ruthlessly describes the sins of a commander, and laid clear his responsibilities for the state in peace and war.

Chapter 6 of the book deals with the remaining chapters (XI, XII and XIII) of *The Art of War*. Sun Tzu insisted on the utmost preparation without which a defeat would be certain. His emphasis on the use of weather and terrain remains strategically important in contemporary era also. Sun Tzu also identified at least six plausible reasons for defeat: wrong assessment of the enemy, greed for power and command, poorly trained armed forces, unnecessary anger by the commanders, lack of discipline in the fighting force, and the poor usage of POWs. He also outlined the significance of intelligence operations and the operators.

In conclusion, this author suggests that politico-military leadership around the world that reads and refers Sun Tzu and takes guidance on policy and strategy, may also adhere to his dicta and strive to resolve conflicts without fighting. The application of hybrid warfare against UMPs should also be discouraged through lawfare and international conventions.

INTRODUCTION

Notes

1. Murat Caliskan and Michel Liégeois, *The Concept of "Hybrid Warfare" Undermines NATO's Strategic Thinking—Insights from Interviews with NATO Officials*, Small Wars & Insurgencies (Taylor & Francis, 2020).
2. Iulian Chifu and Gabriel Anghel, "Hybrid Warfare, Lawfare, and Informational Warfare," in *The Changing Face of Warfare in the 21st Century* (Routledge, 2017), 32.
3. Frank G. Hoffman, *Conflict in the 21st Century: The Rise of Hybrid Wars* (Potomac Institute for Policy Studies, December 2007), accessed August 27, 2022, http://www.potomacinstitute.org/images/stories/publications/potomac_hy bridwar_0108.pdf.
4. Andres B. Munoz Mosquera and Sascha Dov Bachmann, "Lawfare in Hybrid Wars: The 21st Century Warfare," *Journal of International Humanitarian Legal Studies* 7, no. 1 (2016): 63–87, accessed August 28, 2022, https://doi.org/10.1163/18781527-00701008.
5. Lord Jopling, "2018: Countering Russia's Hybrid Threats," accessed January 25, 2023, https://www.nato-pa.int/document/2018-countering-russias-hybrid-threats-jopling-report-166-cds-18-e.
6. Özlem Kayhan Pusane, "How to Profile PYD/YPG as an Actor in the Syrian Civil War: Policy Implications for the Region and Beyond," in *Violent Non-state Actors and the Syrian Civil War*, ed. Ö. Oktav, E. Parlar Dal, and A. Kurşun (Springer, 2018), accessed January 27, 2023, https://doi.org/10.1007/978-3-319-67528-2_4.
7. This author's article, "Future Wars: Dissidence, Despondency & Disappointment (D-3)," *Daily Times*, Pakistan, January 9, 2023.
8. Gary Machado Alexandre Alaphilippe, et al., "Indian Chronicles: Deep Dive into a 15-Year-Old Operation Targeting the EU and the UN to Serve Indian Interests," *EU DisinfoLab*, December 9, 2020, accessed April 8, 2021, https://www.disinfo.eu/publications/indian-chronicles-deep-dive-into-a-15-year-operation-targeting-the-eu-and-un-to-serve-indian-interests/.
9. Donald Stoker and Craig Whiteside, "Blurred Lines: Gray-Zone Conflict and Hybrid War—Two Failures of American Strategic Thinking," *Naval War College Review* 73, no. 1 (2020), Article 4, accessed August 27, 2022, https://digital-commons.usnwc.edu/nwc-review/vol73/iss1/4.
10. Gergely Tóth, "Legal Challenges in Hybrid Warfare Theory and Practice: Is There a Place," in *The Use of Force against Ukraine and International Law: Jus Ad Bellum, Jus In Bello, Jus Post Bellum*, ed. Sergey Sayapin and Evhen Tsybulenko, 18 (Springer, 2018), 173.
11. Sascha-Dominik Bachmann and Hakan Gunneriusson, "Hybrid Wars: The 21st Century's New Threats to Global Peace and Security," *Scientia Militia, South African Journal of Military Studies* 43, no. 1 (2015): 77–98.
12. Vladimir I. Batyuk, "The US Concept and Practice of Hybrid Warfare," *Strategic Analysis* 41, no. 5 (2017): 464–77.
13. Andrew Mumford, "The Role of Counter Terrorism in Hybrid Warfare," report prepared for NATO Centre of Excellence Defence against Terrorism (CEO-DAT), November 2016, accessed August 28, 2022, https://www.europenowjournal.org/2018/11/07/global-hybrid-threats-and-european-security-in-the-age-of-trump-growing-populism-and-international-terrorism/.

INTRODUCTION

14 Dr. Giray Sadik, "Global Hybrid Threats and European Security in the Age of Trump, Growing Populism, and International Terrorism," *EuropeNow Daily*, November 2, 2018, accessed August 27, 2022, https://www.europenowjournal.org/2018/11/07/global-hybrid-threats-and-european-security-in-the-age-of-trump-growing-populism-and-international-terrorism/.

15 Patrick Wintour, "Iran and Saudi Arabia Agree to Restore Ties after China-Brokered Talks," *The Guardian*, March 10, 2023, accessed March 14, 2023, https://www.theguardian.com/world/2023/mar/10/iran-saudi-arabia-agree-restore-ties-china-talks.

16 Zia Ul Haque Shamsi, *Nuclear Deterrence and Conflict Management Between India and Pakistan* (Peter Lang, 2020), 3.

17 This author's article, "WAR: Waste of Available Resources," *Daily Times*, Pakistan, May 17, 2021.

18 This author quotes Andre Gromyko in various articles and his book, *South Asia Needs Hybrid Peace* (Peter Lang, 2021), 16.

19 Jari Tanner, "Finland to Boost Security at Russia Border with Amended Law," *AP News*, July 7, 2022, accessed March 14, 2023, https://apnews.com/article/nato-russia-ukraine-migration-sweden-moscow-002d36695a2e29f1d70d2c85faee225c.

· 1 ·
THE ART AND SCIENCE OF HYBRID WAR

Introduction

The concept of hybrid war is not new, and its tenets include all shades of kinetic and non-kinetic warfare. However, what is new is its scientific application in an artistic manner. The artistic mannerism of hybrid war deals mainly with intangible elements: psycho-social, mind-making, negative perception development and management. The primary objective of the executioner remains to create uncertainty and dissatisfaction about the future of the state among the majority of the population. Whereas, the scientific application of the elements of hybrid war deals with tangible impact on people through economic coercion, cyber-crimes, physical impact, and even territorial capture. Perhaps, this *hybridity* makes hybrid warfare the most favorite tool of application by all strategists and practitioners. Moreover, the intangible applications are relatively less risky and do not lead to activation of any international law or violation of major treaties or agreements, at least for now.[1]

This author argues that the concept of hybrid war and its elements include a host of kinetic and non-kinetic approaches and applications as have been employed in all previous wars and conflicts. This view is shared by most academics who have studied hybrid operations particularly in ongoing European conflicts. "Hybrid warfare is nonetheless an old strategic concept, reminiscent of compound warfare, which consisted of a regular force increasing its operations with irregular means."[2] However, the deployment of its artistic applications using sophisticated technologies has been extremely critical in getting quick results, particularly in narrative building, opinion molding, and perception creation. Raugh is also of the same opinion that "[h]ybrid warfare is not a new concept, but its potential is becoming more sophisticated and deadly and requires novel action."[3] The artistic application of intangible elements of hybrid warfare was witnessed during 2005–20 by Pakistan at the hands of Indian-sponsored campaign, as discovered by European watchdog EU DisinfoLab. Although, Pakistan had been highlighting India's efforts to soften its inner ring at the international forums, these concerns were not given any weightage until the "Indian Chronicles" appeared.[4]

On the other hand, the scientific application of kinetic element is in full swing in Europe where an "Iron Curtain" is being drawn between Russia and Ukraine, as the war between the two neighboring states has entered the second year. While protracted wars are counterproductive, highly destructive, and expensive, yet the end of this war is not in sight.

Art or Science?

Interestingly, the art of executing hybrid war is to evade international obligations and help perpetrators to gainfully employ the intangible elements of warfare. The purpose remains to weaken the target state from within and make people feel insecure, uncertain, and lose confidence in the government and its institutions. Whereas, the scientific execution of hybrid war elements is aimed at physical destruction of the target state through strategic application of sophisticated weaponry and technological means.

The North Atlantic Treaty Organization (NATO), at the organizational level, has been trying to deal with the attacks with the support of non-state and state actors categorized as the hybrid form of warfare. For the purpose, NATO created a comprehensive defense strategy (2010–12). As per NATO's 2010 Bi-Strategic Command Capstone Concept, hybrid and other nonlinear threats have become difficult to address through military action. Moreover, NATO defined hybrid threats as those "posed by opponents with the capability to employ non-conventional and conventional means to pursue their aims simultaneously."[5] However, in 2015, NATO developed a strategy for minimizing hybrid threats through geostrategic environments enhanced by globalization. This policy was seen as a response to Russia's actions in Ukraine and Crimea, for which "Russia utilised military, security, political, technical, legal, intelligence, and economic means to enhance its interests."[6]

Modern-day hybrid wars are perhaps more challenging due to an expanded canvass of operations that combines traditional, irregular, and terror components.

Raugh opines:

> After its wars in Iraq and Afghanistan, the US has increasingly focused on antisocial doctrines. However, insurrection is not the only challenge the United States faces in structuring its armies. Each age exhibits its own type of war, with its limited conditions and its prejudices. It is important that the US and other world powers not focus on post-Cold War insurgency tactics.[7]

The deployment of hybrid warfare tactics is not new in the conceptualization of war.[8] However, certain events across all regions suggest that new practices under the umbrella of hybrid warfare have the potential to alter strategic calculations and achieve the desired impacts. Since the early twenty-first century, military theorists (mostly American) have debated the concept of hybrid warfare, but formal recognition of the concept in military doctrine has yet to be established. Hybrid warfare tactics commonly adopt elements of four existing full-spectrum warfare modes: conventional warfare; irregular warfare (i.e., counterinsurgency and terrorism); related asymmetric warfare (i.e., unconventional warfare as partial warfare); and composite warfare, whereby irregular forces are deployed concurrently against an adversary and used by state actors to complement normal warfare approaches. The entire scheme of contemporary warfare was arrayed in sync with Sun Tzu's concepts of winning the war without fighting.

The most probable driver of the emergence of hybrid threats is grounded in the evolving nature of the global order. "As a denominator, 'hybrid' identifies a combination of battle spaces, types of operations—military or non-kinetic—and a blurring of actors with the scope of achieving strategic objectives by creating exploitable ambiguity."[9] According to Colin S. Gray, since "the character of warfare in a period is shaped, even driven, much more by the political, social, and strategic contexts than it is by changes integral to military science,"[10] the perpetrators of hybrid war seek to alter the character of the target state for which they are willing to go to any length.

Another definition suggests, "Hybrid threats are activities that target the vulnerabilities of opponents as they relate to legislation, history, societal polarization, outmoded practices, ideological differences, technological disadvantages and other geostrategic influences."[11] In fact, the use of non-violent methods is the essence of hybrid warfare and reflects the shift of modern conflict from military to non-military means. Today, international confrontations unfold in physical, information, digital, cultural and cognitive arenas. Increasingly, the tactics used by non-military actors play a significant role in their political and strategic goals, and the efficiency of these tactics has often surpassed those of traditional military means.

The characteristics of hybrid war have manifested as a combination of violent acts with irregular forms of confrontation, such as terrorist activities, cyberattacks, economic and diplomatic sanctions, intelligence sabotage and several other components.[12] As a result, the term *hybrid warfare* has been proposed to encompass the new complex and spatial multidimensionality of these tactics. Moreover, within the framework of hybrid warfare, the open use of military forces often occurs only in the final stages of an attack, and then only under the jurisdiction of the existing international regulatory framework for peacekeeping and crisis management.[13] Moreover, the use of non-violent methods is the essence of hybrid warfare and reflects the shift of modern conflict from military to non-military means. However, this is entirely opposite in case of war against UMPs, unfortunately though.

In 2015, the NATO launched a new hybrid war program prompted by the activities of Russia and Daesh, which altered the balance of Euro-Atlantic security and the stability of the Middle East.[14] Moreover, Russia has used military, legal, political, intelligence, economic and technical elements to advance its hybrid war activities. At the 2014 Wales Summit, NATO adopted its Readiness Action Plan (RAP) to respond to new threats, which was complemented with a hybrid warfare strategy at the Warsaw Summit in 2016. Thus, the question that emerged is how NATO, which is primarily focused on providing common defense for its members, will change its structures and procedures to respond to hybrid warfare challenges. Hence, developing a long-term strategy can benefit NATO allies in opposition to those who use hybrid warfare tactics. However, the same was not done within the specified time and NATO is forced to provide support to Ukraine against Russia in an incoherent and ad hoc manner that may not be sufficient to save the Ukrainian territory from Russian advances.

In fact, in the changed paradigm, a combination of tactics, harmful acts, and attacks within and beyond boundaries can all be considered hybrid threats. However, the term *hybrid threat* has been controversial since it entered the lexicon. According to NATO's definition, "a hybrid threat involves the simultaneous and adaptive use of both conventional and non-conventional tactics to achieve an objective."[15] Other definitions of *hybrid threat* include a novel integration of conventional and non-conventional abilities and unrestricted operational actions.[16] The general characteristics of hybrid threats include criminality, blended modalities, fusion and simultaneity. In its present context, the hybrid threat concept supported by current threat actors defines a new form of warfare using communication networks and technologies, mainly cyber and space.

In fact, since decolonization began, time and again, it has been proven that occupation of captured land is extremely expensive and perhaps counterproductive, no matter how rich the invaded country is in terms of hydrocarbons and natural resources. Afghanistan and Iraq are recent examples. Occupiers may have wanted to stay on ground but eventually had to leave. The noise about hybrid warfare, though the concept is as old as warfare itself, is perhaps to mitigate the exorbitant expenditures on the modern-day military campaigns alongside issues related to lawfare on the human and material losses incurred during violence. Whereas the hybrid warfare that is launched in synergy with media campaign, economic coercion, cyberattacks, espionage, political interference, electoral engineering, and limited military intervention, if required, may be less expensive, and be able to avoid international laws. The purpose would be to initiate Dissidence, Despondency, and Disappointment (D3)[17] among the populace of the target state. By doing so, the perpetrators are able to cause great harm to the target state without raising alarm bells among international institutions, and with considerably reduced expenditures on the war efforts, thereby following the precepts of Sun Tzu.

In addition, hybrid operations are often defined as those that employ military and non-military tools in an integrated and synergetic campaigns aimed at achieving certain objectives, obtaining psychological benefits, leveraging diplomatic actions, employing digital and cyber operations, concealing military and reconnaissance actions and exerting economic pressure on target states.[18] Thus, the term *hybrid warfare* describes a range of threats that differ from conventional threats. In general, one of the most important characteristics of hybrid warfare involves the blurring of legal boundaries by employing destabilizing and subversive measures in such a way that they are undefined as either war or peace. As a result, one of the main objectives of hybrid warfare is to operate outside of the legal boundaries of existing international security organizations. All this is possible due to a factor known as hybrid uncertainty, or the development and implementation of international relations and technologies that can be characterized as treacherous and immoral.[19] Moreover, the term *hybridity* refers not only to combat situations and the conditions, strategy and tactics of rival states but also to the responses that states and their allies create and maintain to combat them. Another conceptual response to this definition of threats is the emergence of the term *irregular military action*, which describes several phenomena, including unconventional and asymmetric military action, airborne sabotage, guerrilla wars, counterinsurgency, uprisings, civil wars and revolutionary actions.

This essential objective of hybrid threats almost breaks the distinction between combatants and citizens while, at the same time, military objectives recede into the background. The tactics used in these types of conflict focus on disinformation and propaganda. They aim to exploit economic, political, technological and diplomatic vulnerabilities; break communities, national parties and electoral systems; and disrupt key infrastructure.[20] Moreover, these attacks are often nonlinear and may have consequences in other spheres. For example, the 2019 drone attacks on oil infrastructure in Saudi Arabia directly affected the global economy.[21]

Though hybridity in threats against an adversary is not a new strategy in non-kinetic warfare, it has certainly paralyzed the international law in its application to avert the threat, particularly against relatively smaller and weaker states. Perhaps, this is one reason that Europe is faced with another war at the moment, which clearly reflects the inability of international laws to avert kinetic and non-kinetic threats of different nature and character.

Interestingly, the art of executing hybrid war is to evade international obligations and help perpetrators to gainfully employ the intangible elements of warfare. The purpose remains to weaken the target state from within and make people feel insecure, uncertain, and lose confidence in the government and its institutions. Whereas, the scientific execution of hybrid war elements are aimed at physical destruction of the target state through strategic application of sophisticated weaponry and technological means.

In fact, the realists' world view manifests wars and conflicts in all its forms in pursuance for power and security by respective states, primarily in their best "national interests." Historically, the most common reasons for wars between states have been for territory, independence, resources, support for allies, etc.

However, the modern wars have had differing causes; rights, freedom, Right to Protect (R2P), preventive, pre-emptive, and against a state to crush the non-state actors (NSAs) of that state. What is interesting is the outcome of any such wars and conflicts, whether in the far past or the near past or even in the ongoing wars and conflicts.

While the US has been engaged in wars and conflicts across the globe, China is expected to fully comply with the precepts of Sun Tzu, and therefore, rely heavily on winning the war without fighting. For the purpose, China is heavily investing in Nepal, Pakistan, Sri Lanka, and a number of African countries, particularly in critical infrastructure like ports, dams and strategic industrial units, to enhance its leverage and gain moral, and diplomatic support, as and when needed.

The scientific application of hybrid threat is seen in the form of cyberattacks, which are concentrated assaults against the digital infrastructure of the target state, including network disruption, malware infection and the use of spam. These attacks do not require the use of force per se, but their level of material destruction to property and loss of life are sometimes comparable to those of armed attacks. While cyberattacks are a *sui generis* war technique that can serve as part of more traditional military campaigns, these are difficult to trace at once, and may take much longer to determine the perpetrators. For example, Russia used cyberattacks as a force multiplier that increased the military capabilities on its armed campaigns against Georgia in 2008 and Ukraine in 2014.[22]

Cyber threats can be divided into four areas.[23] First, *cyber espionage* is deployed in political, economic and military spheres. Many states, notably China, Russia, Iran, and the US, routinely employ cyber-espionage tactics. States typically conduct cyber-espionage activities either directly using their intelligence services or through corporate agents. Second, *cybercrime* is usually conducted for profit, with an estimated effect on the global economy of 2 percent of GDP. Primary cybercrime activities involve information theft, fraud, and money laundering that are typically by terrorist organizations, organized crime and hackers. Third, *cyber terrorism* seeks to obtain a broad range of information from individuals. Here, the main actors include terrorist organizations and state intelligence agencies. Cyber terrorism holds several advantages over conventional terrorism in that it secures a degree of anonymity, provides a greater cost-benefit ratio and has an advantage in terms of delimitation. Finally, *hacktivism* targets digital services and can involve theft and unauthorized publication of information as well as terrorist activities. For example, ISIS uses digital tactics in its recruitment activities. Hacktivism has a specific purpose: to subvert state order and disrupt social peace. This is an emerging high-impact threat. These terminologies are further explained with references in the following chapters.

Conclusion

While the nature and character of warfare keep evolving and so are the strategic applications of the emerging technologies. The sole purpose remains: cause maximum harm to the adversary. However, there is little consideration to what Sun Tzu had said some 2,500 years ago that wars may be won without fighting and if the kinetic applications are essentially required, it must be discreet and for a shortest possible time.

Today's warfare, though not much different in essence, but in approach and applications are hybrid in nature. In order to gain on the adversary in an effective manner, it requires artistic as well as scientific application. Artistic applications mostly deal in the non-kinetic domain, while scientific approach is needed in kinetic applications. While hybrid applications became more popular since the beginning of the twenty-first century, but there is no relief from the physical wars, particularly for the relatively weaker states, though they could be defeated without kinetic applications, following Sun Tzu's precepts.

Notes

1 This author's article, "The Art and Science of Hybrid War," *Daily Times*, Pakistan, July 25, 2022.
2 Chifu and Anghel, "Hybrid Warfare," 32.
3 David L. Raugh, "Is the Hybrid Threat a True Threat?," *Journal of Strategic Security* 9, no. 2 (2016): 1–13.
4 Alexandre Alaphilippe, Adamczyk, and Grégoire, "Indian Chronicles."
5 Iulian Chifu, "Hybrid Warfare, Lawfare, Information Warfare," in *The Changing Face of Warfare in the 21st Century* (Routledge, 2017).
6 Alaa Al-Aridi, "Legal Complexities of Hybrid Threats in the Arctic Region," *Teise/Law* 112 (2019): 107–23.
7 Raugh, "Is the Hybrid Threat a True Threat?," 1–13.
8 Bettina Renz, "Russia and 'Hybrid Warfare,'" *Contemporary Politics* 22, no. 3 (2016): 283–300.
9 Vladimir Rauta, "Towards a Typology of Non-state Actors in 'Hybrid Warfare': Proxy, Auxiliary, Surrogate and Affiliated Forces," *Cambridge Review of International Affairs* 33, no. 6 (2020): 868–87.

10 Colin S. Gray, "The Changing Nature of Warfare?," *Naval War College Review* 49, no. 2 (1996), Article 3, accessed August 28, 2022, https://digital-commons.usnwc.edu/nwc-review/vol49/iss2/3.
11 Bachmann and Gunnariusson, "Hybrid Wars," 77–98.
12 Joseph Dvorak, "Complexity in Modern War: Examining Hybrid War and Future U.S. Security Challenges," MSU graduate theses 3029, 2016. https://bearworks.missouristate.edu/theses/3029.
13 Miroslaw Banasik, "Challenges and Threats for the International Security as the Consequence of the Russian Federation's Hybrid War," *Science & Military Journal* 12, no. 1 (2017): 27.
14 Murat Caliskan and Paul Alexander Cramers, "What Do You Mean by 'Hybrid Warfare'? A Content Analysis on the Media Coverage of Hybrid Warfare Concept." *Horizon Insights* 4 (2018): 23–36.
15 Sascha Dov Bachmann and Andres B. Munoz Mosquera, "Lawfare and Hybrid Warfare—How Russia Is Using the Law as a Weapon," *Amicus Curiae* 102 (2015).
16 Anne Trebilcock, "The ILO as an Actor in International Economic Law: Looking Back, Gazing Ahead," in *European Yearbook of International Economic Law 2019*, European Yearbook of International Economic Law, ed. Marc Bungenberg, Markus Krajewski, Christian J. Tams, Jörg Philipp Terhechte, and Andreas R. Ziegler (Springer, 2019), 1–32.
17 Some of these ideas were published in an Opinion Article, "Future Wars: Dissidence, Despondency, and Disappointment (D3)" by this author in *Daily Times*, Pakistan, on January 9, 2023.
18 O. M. Kikinezhdi and I. M. Shulha, "Challenges of the Hybrid War: Gender in the Mass Media," International Scientific and Practical Conference, Vilnius, August 2019.
19 Oleksandr Moskalenko and Volodymyr Streltsov, "Shaping a 'Hybrid' CFSP to Face 'Hybrid' Security Challenges," *European Foreign Affairs Review* 22, no. 4 (2017): 513–32.
20 Pusane, "How to Profile PYD/YPG as an Actor in the Syrian Civil War," 73–90.
21 Mariam Isa, "How Saudi Oil Attack May Impact SA," *Finweek*, September 26, 2019: 12.
22 Gregory F. Treverton, et al., *Addressing Hybrid Threats* (Swedish Defence University, 2018), 1–93.
23 Nicu Popescu, "Hybrid Tactics: Neither New nor Only Russian," *EUISS Issue Alert* 4, January 2015.

· 2 ·
AVOID PROLONGED WARS

Introduction

This chapter deals with the cardinal principles outlined in chapters I and II of Sun Tzu's book. The most important aspects of chapter I from his book are: Secrecy in Planning, Pretention, Surprise, Utmost Preparation, and Comparative Analysis. Whereas, chapter two of Sun Tzu's book insists on avoidance of a prolonged war alongside his assertions that war is dangerous and expensive and resources would never be adequate for a long-drawn war. Moreover, there will be inflations during wars; therefore, short, swift and quick decision-making will be needed to finish the war early. Sun Tzu insisted that one should only fight to win and for that motivate the troops with rewards. Also treat the POWs humanely so that you could engage them later on to fight for you.

While all the points are extremely important and remain relevant in modern wars even after the passage of 2,500 years, this author intends to discuss an all-important factor of Sun Tzu's insistence on avoidance of a prolonged war. The other important aspects that this author intends to analyze include: war is dangerous, and expensive with high inflation, for which the resources would always be inadequate. However, if war is unavoidable, a short, swift, and quick decision is strongly recommended. Unfortunately, there is no visible effort or the urgency to conclude kinetic operations in any war by the stakeholders.

The Consequences of Ignoring Sun Tzu

There are numerous study centers, think tanks, and departments in every university all over the world, teaching about conflict management, resolution, crisis management, prevention, etc., without any worthwhile contribution in how to avoid modern wars. Nearly 40 percent of the world's population is facing conflicts, wars, or war-like situations with no solution in sight. The stronger nations do not think twice before initiating a war against an UMP, at times, without any consideration of the outcome. The Iraq War (2003–11) and the war in Afghanistan (2001–21) are examples of twenty-first-century wars, which were grossly mismanaged and continued for decades without any logic.

Perhaps, it is time to insist on the implementation of Sun Tzu's precepts that if kinetic application is extremely necessary, than it must be discreet and for the shortest possible time. Also, it is extremely important that we start to invest in "War Management Studies" instead of "Conflict Management" or "Crisis Management," because it is the mismanagement of conflicts and crises that leads to wars, and it is the mismanagement of wars that is causing the death and destruction of civilians and non-combatants. Therefore, it is necessary to study war management as a subject of Social Sciences and analyze the impacts of often irrational and ill-considered decisions by policymakers to wage wars without a well-thought-out minimum-damage strategy and a sound exit strategy.[1]

Let's have a look at the two decades-long Afghan war that started in the aftermath of the tragic incidents of 9/11. Initially, the objectives were to eliminate al-Qaeda from Afghanistan to ensure that Global War on Terror was won, and there was no danger of terror activities, particularly against the US. However, the war continued even though al-Qaeda was neutralized much earlier than anticipated by the US. Two decades later, President Biden withdrew from Afghanistan after the US had signed an Agreement with the Taliban in Doha-Qatar on February 29, 2020, under the Trump Administration.

The post-9/11 Afghan war was grossly mismanaged in terms of its initiation, execution, continuation, and even culmination. The US failed to evaluate the resolve of the Afghan people in that they did not accept the foreign occupation and had defeated two global powers of the time earlier: Great Britain and the USSR. The continued war did not have well-thought or clearly defined political or military objectives that were doable or preferable. The fight continued for years and years against the same ill-equipped but determined fighters until the US and NATO decided to call it a day. Likewise, the hasty withdrawal of the world's best-armed forces demonstrated clearly that the occupation was without a purpose and unethical.

The ongoing Yemen War also presents a similar picture. The primary objective of reinstalling the Hadi government was quickly achieved, yet the war is continuing in its ninth year, with the worst famine and gross human rights violations.

Sun Tzu insisted that one should aim only for victory, and motivate the troops against enemy and reward them after victory. He also laid great emphasis on treating POW, humanely.

This chapter is particularly aimed at highlighting the significance of Sun Tzu's specific dictum about protracted wars and conflicts. This dictum has been violated by several states at different times in history with similar results: destruction, stalemate, ruined economies, and prolonged human sufferings.

The twentieth century is replete with such wars and conflicts starting with World War I (1914–18), after which nearly the whole of Europe needed to be rebuilt. The Treaty of Versailles signed between the victors and losers was meant to ensure that the Germans would never be able to stand on their feet, and therefore, the world would be a safer place to live. The formation of *League of Nations* on January 10, 1920, was also for the same purpose. However, a resurgent Germany, under Adolf Hitler, was ready to storm Europe within two decades.

The twenty-first century is no different and in little over two decades, a number of regions have seen wars and conflicts. Some of these wars can be categorized as long wars: Gulf War-II, Afghan War-II, Yemen War, Syrian conflict, and no letup in Arab-Israel conflict and India-Pakistan enduring rivalry.

Prolonged wars, as prophesied by Sun Tzu, bring no benefit to any of the stakeholders. History is replete with examples of wars and conflicts which lasted for years. The erstwhile Soviet Union invaded Afghanistan on December 24, 1979, and remained there until February 15, 1989. In the bargain, the Soviet Union ended up losing its political identity and disintegrated for good, perhaps because the invasion was seen as a blatant violation of Afghanistan's sovereignty. Nearly all nations joined hands led by the US, and its allies, and supported the Afghan *Mujahedeen's* resistance. Pakistan not only played the role of a frontline state against Soviet occupation but also became the training and logistics lifeline of the freedom struggle.

Soviet leadership of the time blatantly ignored the golden rule of Sun Tzu to avoid a prolonged military campaign and paid the price that not only changed its political identity but also the international system. The bilateralism was shattered and the emergence of new world order under unilateralism gave the US a free hand to become a global policeman.

Perhaps, the history repeated itself and the US entered Afghanistan in 2001 following the 9/11 attacks to fight the Global War on Terror (GWOT). This entry was supported and legitimized by global powers and institutions. The US and NATO forces stayed in Afghanistan with full military presence for over 20 years and left on August 15, 2021, only to leave the war-ravaged country again to the same government led by Taliban, which was not recognized by the US-led Western world at the time of invasion, and is still not recognized by the international community.

While the US was still looking for an excuse on its failures in Afghanistan, Russia decided to enter Ukraine, in an effort to extend its perimeters to block NATO forces on its doorsteps. For years, the Kremlin had been pushing against NATO's expansion, particularly against Ukraine, because its joining the military alliance was seen as a genuine security threat. It is evident now that Ukraine was Russia's red line. However, without going too far back into history, this chapter will only review the contemporary wars that can be categorized as the protracted conflicts. The ongoing war between Russia and Ukraine, will also be studied because it has the potential to expand horizontally as well as vertically.

While recognizing the genuine security concerns of Russia due to NATO's eastward expansion, Pakistan's former Chief of Army Staff (COAS) General Qamar Javed Bajwa did not endorse Russia's Ukrainian campaign. He was speaking at the Second Islamabad Security Dialogue on April 2, 2022. General Bajwa categorically stated that Pakistan is against "camp politics" and urged the Russian government to cease hostilities to save precious lives.

While NATO has been expanding its frontiers since the demise of the erstwhile Soviet Union on December 31, 1991, successive Russian leadership had been warning about consequences if its "red line" was crossed. Since Ukraine was Russia's red line and as the narrative goes, Ukrainian leadership was keen to join the European Union (EU) as well as NATO, the Kremlin could not remain silent and ordered its forces to cross the international border on February 24, 2022.

The Russia-Ukraine war is now into the second year, and cannot be categorized a protracted war as yet. However, the ongoing hostilities are a consequence of the protracted conflict between Cold War rivals. NATO countries are providing arms, equipment, and financial support to Ukraine and in a way preparing it to oppose Russia's insistence that Kyiv must not join NATO due to its legitimate security concerns.

Ukraine, though much smaller and relatively weaker than its much larger and militarily stronger opponent Russia, has been able to withstand the pressures of Russian forces with more advanced weapon systems provided by her Western neighbors. This particular aspect was highlighted by Pakistan's COAS General Bajwa as well that the lesson one can draw from the ongoing war is that relatively weaker armed forces can put up a reasonable resistance with modern weapons. He called for the need to modernize the armed forces with state-of-the-art equipment to counter the numerical strengths of the adversary in an evolving geostrategic environment.

On the other hand, Russian campaign has not progressed as planned. Russia's foremost political objective was to ensure that Ukraine accepts the latter's supremacy in the region and the condition of not joining NATO in times to come. Moreover, Kyiv must not resist Russian forces ingress and accept its prescription of peace in the region. Hence, Moscow's campaign has started to expand and is not likely to end anytime soon due to Western support for the Ukrainian resistance.

While Russian security concerns are well supported, the probability of resolving the issue through a short and swift kinetic action seems a far cry. Russia seems to have violated Sun Tzu's precepts of avoiding a prolonged war to achieve certain political objectives. With every passing day on the battlefield, global anger against Moscow will increase, for which US-led Western powers are constantly working. Images of dead bodies and destroyed infrastructure on social media would further raise concerns among the neutrals. Moreover, the harsh sanctions imposed on Russia by the US-led Western powers would further lead to increase in energy prices and hurt majority of countries in Europe, Asia, and Africa, besides the country's own economy.

The Russia-Ukraine war has already caused a large-scale devastation and casualties. The Western world is supporting Ukraine morally, militarily, and financially. On the other hand, the global economy has been hit hard due to this new war in Europe, primarily because Russia controls energy supplies to a number of European countries. The country remains the biggest exporter of crude oil and earns nearly USD 123 billion a year. Moreover, it is not only a major supplier of just oil and gas, but also wheat, metals, and fertilizers, as well.

The US and other European partners have imposed compelling sanctions on Russia and its major exporting companies due to its imposed war on Ukraine. This has had serious implications for the Russian economy, and hence forced President Putin to insist on receiving Russian Rubles in exchange for gas exports to European countries.

Moreover, both Russia and Ukraine are the biggest exporters of food grain to a number of countries across the globe. For instance, Egypt was the biggest importer of Ukrainian wheat last year alongside Lebanon, and Pakistan. Ukraine produces about 7 percent of the world's wheat as well as sunflowers, corn, soybeans, and barley in large exportable quantities, particularly to North African countries.

Historically, a prolonged military campaign usually proves counterproductive for the initiator. Vietnam War for the US, first Afghan war for the Soviets, and second Afghan war for the US are ample proof of this argument. Russia must not lose sight of history and avoid prolonging its Ukrainian campaign before it proves to be another Afghanistan for the global power.

Following Sun Tzu, Russia must make an effort to quickly achieve its politico-military objectives in Ukraine and avoid getting into the syndrome of mission creep, which may raise the likelihood of a larger conflict in the region.

Dilating upon the essentials mentioned by Sun Tzu in Chapter I of the *Art of War*, each tenant appears to be extremely beneficial for the perpetrators of hybrid war. The Secrecy in Planning, Pretention, and Surprise are interdependent and interlinked in any hybrid campaign primarily because one guarantees the success of the other to achieve the cumulative effect.

Pakistan army, in order to retake the Siachen glacier that was illegally occupied by India in 1983, planned a military operation on Kargil Heights in 1999 that could oversee the Kargil-Leh Highway (NH-1), to choke India's critical supply lines to Siachen. The hybrid mannerism of campaign planning ensured extreme secrecy, and even the major stakeholders, including the Pakistan Air Force (PAF) and Pakistan's Foreign Office, were not fully aware of the plan. In fact, India's Border Security Forces (BSF) have their posts on Kargil ridges but they vacate these in winter months due to extreme weather conditions, assuming that Pakistan also cannot occupy them in these months. Moreover, Sun Tzu's advice on pretention was also skillfully deployed and people sent across the Line of Control (LoC) were disguised as the locals.

Through the adherence to secrecy in planning and pretention, Pakistan Army achieved a perfect surprise and occupied the Kargil Heights without any resistance, rather without any knowledge of Indian BSF, until it was very late and Indian Air Force (IAF) had to be scrambled to get the area cleared by force, for the first time since the 1971 war.

PAF was not part of the campaign plan and not called to contest IAF, perhaps because a direct military clash between the two air forces was to be averted. However, the two land forces did clash fiercely, ringing alarm bells in the world capitals. The two neighboring states: India and Pakistan had declared themselves as nuclear states only a year ago and it was highly unusual in the post-nuclearized world that states with nuclear status clashed militarily. Historically only once have the two nuclear states had clashed briefly when China and the Soviet Union's border forces undertook limited skirmishes along the Ussuri River in 1969. However, a far serious situation took place in October 1962 when the two super powers of the Cold War era: The US and the USSR, came close to a nuclear conflagration.

Soviet Union secretly placed Medium Range Ballistic Missiles (MRBMs) at the door steps of the US: Cuba. The action was seen as a violation of Monroe Doctrine, which was coined by President James Monroe in 1823, and called for non-interference in Western Hemisphere.[2] The October 1962 events commonly referred as the Cuban Missile Crisis are still regarded as the most dangerous situation of the Cold War when the world was only a few steps away from a nuclear war. Fifty years on, the Cuban Missile Crisis are still remembered with pain and anguish, and is perhaps the most researched and most referred events of the Cold War. President Biden also made a mention in the context of Ukraine War while responding to President Putin's nuclear posturing statement. "For the first time since the Cuban Missile Crisis, we have a direct threat to the use of nuclear weapons, if in fact things continue down the path they'd been going, ... We have not faced the prospect of Armageddon since Kennedy and the Cuban Missile Crisis."[3]

Back to Kargil and the Sun Tzu's precepts on Secrecy in planning, Pretention and the Surprise, unfortunately the planners only remembered these few dicta and ignored the rest, and hence faced an embarrassing fallout when the US President Clinton forced Pakistan's Prime Minister Sharif to withdraw from the occupied ridges. Perhaps, the tactical superiority achieved within the zone of operation was lost quickly due to the strategic neglect of Sun Tzu's another dicta in the same chapter that an utmost preparation is needed if war is extremely necessary and must be planned to the minutest of details.

Another ongoing war that needs attention due to its longevity is Yemen War. Notwithstanding the historical instability in Yemen, the ongoing military conflict between Saudi-led Gulf Cooperation Council forces and perceived Iranian-backed Houthis has further fueled the age-old animosity between Saudi Arabia and Iran, based on the Sectarian divide. The expanded Iranian influence in the Arab world starting from Iraq, Syria, Lebanon, and to Yemen, perhaps left Saudi Arabia with no other option but to form a military alliance to deny a foothold on the Yemen's territory to break Iran's ambition of strategic strangulation of the GCC. Hence, the military response by Saudi-led forces to Houthis' expanding control on the territory was not only a political option but a strategic compulsion of the GCC, particularly Saudi Arabia and UAE.

The execution of Saudi-led military operations in Yemen was flawed right from the beginning, perhaps because not much effort was made to secure peace and calm through negotiations. Houthi fighters took control of the capital, Sanaa, in September 2014, and pushed toward the Southern port city of Aden. The situation kept on deteriorating due to which President Hadi had to fly out of Sanaa. Much later in March 2015, a coalition of Arab states led by Saudi Arabia launched a military campaign to drive out the Houthis from Sanaa to reinstall the President. This is perhaps the only political objective that the Saudi-led coalition has achieved; the reinstallation of Abd-Rabbu Mansour Hadi as the internationally recognized president of Yemen. However, the same has divided the country and its populace to continue a bitter civil war, which has created a humanitarian crisis in the Yemen. Shireen Al-Adeimi concludes after four years of Saudi-led military operations in Yemen that "this brutal and ongoing onslaught has taken the lives of more than 60,000 Yemenis and left half the population—14 million people—on the verge of famine."[4] She quotes United Nations and categorically declares the crises in Yemen as the world's worst humanitarian crisis.[5]

The political decision was made (on the execution of the operation Al-Hazem and the subsequent campaign of Restoration of Hope) based on the stated reason to restore the legitimacy of President Abdrabbuh Mansur Hadi, but the undeclared reason was to stop the Iranian expansion, especially its support in arming and training Houthis, as well as the Shiite domination over the Kingdom of Bahrain, Syria and Lebanese Hezbollah, in view of the fact that Saudi Arabia felt that it is surrounded by many from the north, especially

Iraq and now from the south as well (Yemen). Perhaps, this was the real political objective behind this operation; however, the decision to launch the operation was taken in a haste and in total disregard to Sun Tzu's dicta of utmost preparation. The result was obvious and now the KSA-led coalition is unable to pull out from the war they initiated on the assumption that it will be finished in weeks.

Another important aspect that Sun Tzu insists deals with treatment of the POWs. Article 13 of Geneva Convention reads, "Prisoners of war must at all times be humanely treated. Any unlawful act or omission by the Detaining Power causing death or seriously endangering the health of a prisoner of war in its custody is prohibited, and will be regarded as a serious breach of the present Convention."[6] Sun Tzu laid great emphasis on this aspect, primarily with a purpose to keep POWs engaged, and impress upon treating them with respect so that they may join you in your future campaigns. This was the norm in the ancient times that the POWs could be reenrolled into the victorious forces.

Conclusion

To conclude this chapter, it is once again reiterated that protracted military engagements are cost prohibitive, highly destructive, with lot of fatalities on both sides. Moreover, it is highly unpopular at home and abroad, and it carries the dangers of horizontal and vertical expansions throughout the period of violent or non-violent engagement. Sun Tzu was of the view that war should only be waged for victory and hence the best strategy should be to win a war without fighting, meaning by the deployment of non-kinetic means be given precedence over the kinetic applications as part of the hybrid warfare package.

For the powerful nations, war continues to be the most favorite option for resolving the disputes or conflicts, particularly against the UMPs. The powerful states operating within the realists' paradigm, are making use of all the elements of hybrid warfare, yet giving priority to kinetic operations instead of deploying them as the last resort. Moreover, in order to benefit their Military Industrial Complex (MICs), the powerful nations prefer to go for a long-drawn war, particularly against the UMPs, so that the smaller and weaker countries are totally destroyed and in the bargain large scale of arms and equipment are expanded. Iraq, Libya, Syria, and Afghanistan are only a few examples where the emerging technologies were tested, sold, and employed by the powerful militaries against inconsequential opponents.

Notes

1. This author's article "Mismanaging Modern Wars," *Daily Times*, Pakistan, March 6, 2023.
2. Monroe Doctrine serves three main concepts for the enduring US interests at the time and perhaps even now—separate spheres of influence for the Americas and Europe, non-colonization, and non-intervention—were designed to signify a clear break between the New World and the autocratic realm of Europe. Accessed November 17, 2022, https://history.state.gov/milestones/1801-1829/monroe.
3. Nathan Williams reports for BBC on October 7, 2022, "Ukraine War: Biden Says Nuclear Risk Highest since 1962 Cuban Missile Crisis of 1962," accessed January 3, 2023, https://www.bbc.com/news/world-us-canada-63167947.
4. Dated March 2019, accessed April 2019, http://inthesetimes.com/article/21806/yemen-war-saudi-arabia-uae-trump-obama-famine-power-khanna-sanders.
5. Ibid.
6. UN Human Rights Instruments, "Geneva Convention Relative to the Treatment of Prisoners of War," Adopted on August 12, 1949, accessed January 4, 2022, https://www.ohchr.org/en/instruments-mechanisms/instruments/geneva-convention-relative-treatment-prisoners-war#:~:text=Prisoners%20of%20war%20must%20at,breach%20of%20the%20present%20Convention.

· 3 ·
KNOW YOUR ENEMY AND KNOW YOURSELF

Introduction

This chapter analyzes chapters III, IV, and V, of *The Art of War*. These chapters are significant due to the contents related to war itself. Sun Tzu had long insisted on wining without fighting and if necessary a short, and fierce attack for a quick victory. In the domain of hybrid applications, he insisted on striving for the psychological ascendency, and avoid killing the enemy soldiers, rather desired on seeking surrender and include them in one's own force.

Sun Tzu laid great emphasis on preparation and surprise, expanding the capacity to fight, and above all the non-interference from political leadership. Sun Tzu's famous dictum that is widely quoted, "know your enemy and know yourself," would also be analyzed through hybrid applications in contemporary wars and conflicts. Since Sun Tzu was against prolonged wars, he vehemently opposed the siege and declared it as a bad strategy. Whereas in chapter IV, Sun Tzu insisted on the soundness of the plan that is offensive, multi-directional, and hence unpredictable. In chapter V, Sun Tzu reiterated on the Commanders to be flexible in their approaches and directions of attacks, and must hide their intentions while pretending to be weak. Interestingly, he prophesied pretention, perhaps opposite to what deterrence demands; strong show of intent.

Know the Unknown Enemy also

The theory of classical realism asserts on the significance of power and security for the state in the prevalent international system. However, these precepts have made the developed states very selfish and cruel. The relatively stronger states have totally disregarded the basic needs of the smaller states and in the garb of own security, have ruined the weaker states through both: kinetic and non-kinetic means. This has made the world highly vulnerable to wars and conflicts, and the twenty-first century has already seen a number of wars between UMPs. The Afghan War-II, Iraq War-II, Libyan War, Yemen War, Syrian Conflict, and the ongoing Russia-Ukraine war, to name the few.[1]

The war in Europe between Russia and Ukraine has entered into its second year since a full-scale offensive was launched by Russian land forces on February 24, 2022. Like so many wars, which begin as a result of miscalculation, or the exuberance of leadership without due consideration, this war is no different. Neither in its nature nor its character. It is intense, destructive, and without a visible end.

In fact, wars do not resolve disputes on permanent basis rather sow the seeds for the next war. Recounting Sun Tzu's precepts that a stronger power must aim to win the war without fighting, or at best win the war in minimum possible time, because the protracted wars would always favor the defender. Moreover, the people would soon get alienated with offensive forces and take up arms to engage the enemy alongside their regular forces, which could be extremely harmful for the occupying forces. The same can be seen in the ongoing Russia-Ukraine war as the world in general sympathizes and supports Ukraine in its efforts to defend its territory. However, the ongoing war in Europe calls for a greater introspection of this paradoxical linkage between security and economy. Ukraine, a developing country by European standards, has vast natural resources, and second largest reserves of natural gas, after Russia. Moreover, Ukraine "accounts for 10% of the world wheat market, 15% of the corn market, and 13% of the barley market. With more than 50% of world trade, it is also the main player on the sunflower oil market."[2]

Ukraine may have been a poor country by European standards, but neither its economy nor military was so weak that any of its neighbors, except

Russia, could launch an all-out military campaign at will. Russia has not only invaded Ukraine, but has started slicing its eastern territories and so far, has declared at least four Russian ethnic regions as autonomous, meaning that it has no intention to leave unless it gets defeated, which is highly unlikely. Russia must understand that the US-led NATO does not want to end this war and would like to keep their enemy (Russia) engaged, perhaps in a similar way as was done in first Afghan war (1979–89).

Inspired by the dictum of great Quaid, the Founder of Pakistan, Muhammad Ali Jinnah about "Peace within and Peace without," this author proposed the phrase "Enemy within and Enemy without," particularly in the era of hybrid war or fifth Generation Warfare or Non-Kinetic Warfare (NKW), where the enemy remains invisible and unrecognizable. Moreover, the political environment is molded in a manner that it becomes exceedingly difficult to distinguish the enemy.

In fact, this is the most desirable situation for an adversary against a target state when its own people are restless and disillusioned. If an enemy's purpose of inducing dissatisfaction among the masses is achieved, it has already won half the battle. Perhaps, the enemy has accomplished what Chinese sage Sun Tzu had professed more than 2,500 years ago that the best victory is one in which you win without fighting.

Sun Tzu had all along insisted on greater knowledge about the enemy but equally important was to carry out a dispassionate self-analysis to avoid being dragged into wrong wars. He had said, "Know the enemy and know yourself; in a hundred battles you will never be in peril."[3] Sun Tzu stated that it is essential to carry out an in-depth analysis of enemy's capability and intent before making a move on the battlefront. For the purpose he laid great emphasis on the organizations of intelligence departments. According to Sun Tzu, it is incumbent upon the military commander to have complete information about enemy's possible plans so that he is not surprised during war. However, with more than eight years into the war in Yemen, it is evident that GCC alliance made wrong assumptions about the strengths, and resolve of the Houthis fighters. The GCC alliance made gross miscalculation about their adversaries' supply chain management of arms and military equipment, allegedly from Iran. Houthis fighters keep on surprising the Saudi leadership with surface-launched missiles on Kingdom's population centers and critical infrastructure. In fact, knowing the enemy becomes extremely challenging if the resistance comes from the NSAs, like in the case of Yemen. This particular aspect related to knowing the enemy well is especially significant in the hybrid warfare scenario.

In South Asia as well, Pakistan has had multiple wars, conflicts and crises with its arch-rival, India. The core issue of most wars and conflicts was Jammu and Kashmir (J&K), but the agenda items on the disputes have proliferated over the past seven decades. While J&K remains unresolved, Sir Creek, Siachen, water sharing, the Kargil Conflict (1999), Twin Peak crises of 2001–2, Mumbai attacks (2008), and lately the February 2019 crisis, to mention a few, have been added to the long list of evolving disputes.

Pakistan has been subjected to an unending hybrid warfare by India to weaken the state from within, particularly since the overt nuclearization of the region in 1998. India has used, perhaps all of the tools against Pakistan, deployed in the process of any hybrid war. The same has been verified by the European watchdog DisInfoLab in the "Indian Chronicles" report released in December 2020. The fifteen-year-long operations were aimed at major capitals in the EU and the United Nations (UN) to serve Indian interests, and bring harm to China and Pakistan. However, India could not have achieved its political objectives without an active support from within the targeted areas. While few external agents were also sent as facilitators and the directors of executing sensitive operations, like Indian Navy's serving officer *Kulbhushan Jadhev*, who was taken into custody from near the Iranian border on March 3, 2016. Jadhev is now in a Pakistani jail under a death sentence, while the Indian High Commission refuses to appoint a defense counsel for his case currently pending at the Islamabad High Court (IHC), following an order for fair trial by the International Court of Justice (ICJ).

Likewise, political crisis leading to socioeconomic instability cannot be triggered without active support from local elements. These may be Non-Governmental Organizations (NGOs), political offices of external elements, ingress through Joint Ventures (JVs), foreign-funded media houses, etc., as was evident from the "Indian Chronicles" report. At least 750 fake media and 550 website domains participated in the campaign directed mainly against China and Pakistan at the behest of Indian intelligence agencies.

Recently triggered political instability in Pakistan which ultimately led to economic downturn is also attributed to externally driven but locally executed move. What the enemies of Pakistan gained from the prevalent situation is skyrocketing inflation, a dissatisfied populace, falling currency, rising dependency on international donor agencies, and an extremely vulnerable society which has become polarized, volatile, and disenchanted with state institutions.

In fact, this is the most desirable situation for an adversary against a target state when its own people are restless and disillusioned. If an enemy's purpose of inducing dissatisfaction among the masses is achieved, then it has already won half the battle. Perhaps, the enemy has accomplished what Chinese sage Sun Tzu had professed more than 2,500 years ago that the best victory is one in which you win without fighting. Therefore, it is extremely important to understand that invisible elements are identified from within because the outside enemy is visible, whereas the enemy within remains invisible and causes more harm than the visible enemy, who can be tackled by the trained armed forces. Pakistan has the experience of tackling the ongoing hybrid war imposed on it by India; therefore, it is essential that more emphasis is placed on identifying the enemy within because outside enemy cannot succeed in its political objectives without active support from elements located inside the target state.[4]

Fierce Attack and Quick Victory

This particular dicta by Sun Tzu's comes into play when the aggressor has exhausted all the avenues of winning the war without fighting and now the kinetic action is considered extremely vital. The attacker has done his homework on planning, execution, and exit strategies in the minutest of details, because he has to know itself and the enemy before any type of military operations is undertaken, no matter how small it is. Moreover, the attacking commander has done his calculations on the force ratio, studied about the direction of attack, terrain and weather en-route and zone of operations, etc. Once the political leadership has given a go-ahead, it is the responsibility of the military commander to ensure that he plans an attack that is fierce and achieves a quick victory without spending too much effort and time in the combat.

However, none of the twenty-first-century wars, even against the UMPs, could be declared as short and swift on the criterion laid down by Sun Tzu. The Saudi leader King Salman, who ordered the launch of Operation Decisive Storm on March 25, 2015, kept the politico-military objectives open that "defending the legitimate government in Yemen, And saving the Yemeni people from Houthi aggression."[5] At the time of its launch, the operation was widely supported by Egypt, Bahrain, Kuwait, Qatar, the United Arab Emirates, Jordan, Sudan, and Morocco. The US, the United Kingdom, and France also committed to provide diplomatic and logistics support during the military operations. However, Qatar was forced by Saudi Arab to withdraw from the coalition following the economic blockade of the country by Saudi-led quartet on June 5, 2017.[6]

Analyzing King Salman's stated goal through the lens of Sun Tzu, one can only predict that it would be a long-drawn war, because the enemy was not visible, and its capabilities were not ascertained properly. Moreover, its logical end perhaps lied in a negotiated settlement and not through kinetic means, thereby implying that deployment of hybrid means could have been more successful than an all-out military operations, which has now entered its eighth year with fears of famine and rising poverty in Yemen.

Likewise, the US revenge against Taliban for 9/11 tragic incidents that occurred due to deliberate banging of two passenger airplanes to World Trade Centers in New York, another on Pentagon Headquarters in Arlington, near Washington D.C., and a fourth plane was forced in a field in Shanksville, Pennsylvania. The tragic attacks killed nearly 3,000 innocent people, and the Bush Administration was well within its rights to initiate a military action against the perpetrators of the attacks. However, without waiting for thorough investigations, and conclusive evidence, the US started a bombing campaign against Afghan soil, once Taliban failed to fulfill the demands of Osama Bin Laden's expulsion from Afghanistan.

The US ignored not only Sun Tzu's precepts on planning, but also about not knowing the enemy. The US should have studied the history of Afghanistan more carefully and known that Afghans do not accept invaders on their soil and have already defeated the British and the Soviets[7] in their endeavors to concur the land. The US should have known that one can occupy Afghanistan by force but cannot capture the country due to unwavering resolve of the people to fight the invaders till the last Afghan is alive. The result was no different and after two decades-long campaign, the US handed over the country back to the same Taliban after the Doha Agreement was signed on February 29, 2020.

The Russo-Ukrainian war has now entered into the second year, with large-scale devastation and casualties. The Western world is supporting Ukraine morally, militarily, and financially. On the other hand, the global economy has been hit hard due to this new war in Europe, primarily because Russia controls energy supplies to a number of European countries.

The US and other European partners have imposed sanctions on Russia and its major exporting companies due to its imposed war on Ukraine. This has had serious implications for the Russian economy, and hence forced President Putin to insist on receiving Russian Rubles in exchange for gas exports to European countries. After initial reluctance, "most of the gas importers have already opened their account in Rubles with Gazprom." "Germany is also a top gas importer and had already paid in Rubles. Like Italy, Germany is a massive consumer of Russian gas," according to a recent Bloomberg report. Another report suggests that "a total of twenty European companies have opened accounts, with another 14 clients asking for the paperwork needed to set them up." Moreover, both Russia and Ukraine are the biggest exporters of food grain to a number of countries across the globe. For instance, Egypt was the biggest importer of Ukrainian wheat last year alongside Lebanon, and Pakistan. Ukraine produces about 7 percent of the world's wheat as well as sunflowers, corn, soybeans, and barley in large exportable quantities, particularly to North African countries.[8]

Likewise, Russia produces and exports massive quantities of wheat, sugar beets, milk, potatoes, cereals, and chicken. Moreover, several European countries entirely rely on gas from Russia, including North Macedonia, Moldova, Belarus, Norway, Serbia, and Bosnia and Herzegovina.

According to the Observatory of Economic Complexity (OEC) report, "the top exports of Russia are Crude Petroleum ($74.4B), Refined Petroleum ($48B), Petroleum Gas ($19.7B), Gold ($18.7B), and Coal Briquettes ($14.5B), exporting mostly to China ($49.3B), United Kingdom ($25.3B), Netherlands ($22.5B), Belarus ($15.8B), and Germany ($14.2B)." The BBC reports: "In 2019, Russia accounted for 41% of the EU's natural gas imports."

However, the Ukrainian war has again brought in the kinetic application to the fore and NATO's assistance to Ukraine do not seem to deter Russia and its offensive continues into the second year of war. International watchdogs are seeing the destruction of Ukraine from a distance with occasional statements and some military support from NATO countries which initially instigated Ukrainian to cross Russian redlines. However, this war in Europe has seriously challenged the deterring capacity of NATO to Russia, and hence placed the dream of a European peace in dire situation.

Conclusion

To conclude this important chapter which dealt with Sun Tzu's famous quote of knowing the enemy and knowing yourself to ensure victory in every military engagement. This author would insist on the importance to understand that at first the enemy within is identified because the outside enemy is visible, whereas the enemy within remains invisible and causes more harm than the visible enemy, who can be tackled by the trained armed forces. Therefore, it is essential that more emphasis is placed on identifying the enemy within because the outside enemy cannot succeed in its political objectives without active support from elements located inside the target state.

Moreover, President Putin must remain cautious of mission creep, and must not violate Chinese sage Sun Tzu's landmark dictum that protracted wars are counterproductive for the offensive elements. Russia may have achieved its politico-military objectives partially and may be willing to continue its war for another two years to slice as much Ukrainian territory as possible. However, it would soon realize that going into the Ukrainian heartland beyond the areas inhabited by ethnic Russians would become improbable to hold without serious repercussions. Putin must not ignore deep sociocultural and family ties with the Ukrainians, and therefore, must bring this war to an end as soon as its own security parameters are achieved.

Notes

1. This author's article, "The Need for an International Security Agreement (ISA)," *Daily Times*, Pakistan, January 16, 2023.
2. Ines Eisele, "Five Facts on Grain and the War in Ukraine," *Deutsche Welle*, January 11, 2022, accessed March 8, 2023, https://www.dw.com/en/five-facts-on-grain-and-the-war-in-ukraine/a-62601467.
3. S. B. Griffith, *Sun Tzu's Art of War* (Oxford University Press, 1971), 84.
4. Some of these ideas were published in an Opinion Article, "Identifying the Enemy within and Enemy without" by this author in *Daily Times*, Pakistan, on June 27, 2022.
5. May Darwich, "The Saudi Intervention in Yemen: Struggling for Status," *Insight Turkey* no. 20, (Spring 2018), accessed January 11, 2023, https://www.insightturkey.com/articles/the-saudi-intervention-in-yemen-struggling-for-status.
6. Qatar Crisis Complicates Yemen Civil War," *Economist Intelligence*, July 12, 2017, accessed January 11, 2023, https://country.eiu.com/article.aspx?articleid=1745662758&Country=Yemen&topic=Politics.

7 The erstwhile Soviet Union invaded Afghanistan on December 24, 1979, and remained there until February 15, 1989. In the bargain, Soviet Union ended up losing its political identity and disintegrated for good, perhaps because the invasion was seen as a blatant violation of Afghanistan's sovereignty.
8 This author's article, "Is Ukraine Becoming Europe's Afghanistan?," *Daily Times*, Pakistan, May 23, 2022.

· 4 ·
SIGNIFICANCE OF CIVIL-MILITARY RELATIONS

Introduction

This chapter analyzes the contents of chapters VI, VII, and VIII from Sun Tzu's book. Sun Tzu demanded creativity in strategy by the commanders. He was a proponent of the delegation of power and authority, and formation of different commands. However, he insisted on keeping enemy engaged and continue relentlessly planning of attacks secretly.

Sun Tzu talks of the policy matters between various organs of the state. He warned that if the state organs are not on the same page, forget about war, and if the armed forces are not one, you cannot win a war. In chapter VIII, Sun Tzu ruthlessly describes the sins of a commander, and laid clear his responsibilities for the state in peace and war.

Sun Tzu laid the foundations of clear delineation between policy and strategy. Unfortunately, the use of these terms interchangeably creates confusion and diminishes the difference in essence and context.

CHAPTER 4

Policy, Strategy, and Doctrine

The age-old debate on the differences between *Policy–Strategy* and *Doctrine* remains relevant even today because each state prefers the use of terminology as per its academic, and organizational requirements, particularly in affairs related to national security. The intimate relationship between *Policy* and *Strategy*, as evident in the strategic literature is perhaps responsible for these terms to be used interchangeably; wrongly though. Moreover, the term *Doctrine* is also used loosely, particularly while discussing it alongside *Policy* and *Strategy*.

The purpose of this debate in this book is to highlight the differences between *Policy–Strategy* and *Doctrine* in essence and contextual domain so that Sun Tzu's precepts on the relationship between the Emperor and the Commander are understood in letter and spirit. Moreover, the explanations of the core differences between *Policy–Strategy* and *Doctrine* would facilitate the researchers to focus on these terminologies independently and not interchangeably. Brief definitions and explanations of relevant terminologies are placed in Appendix A. These definitions are taken from the published sources, however, at places the reference may be missing.

In fact, the term *Policy* takes its inspiration from a French term *Policie* from the sixteenth century. The policy defines ends which are in conformity to the purpose of the state. The policy caters for all possibilities for the defined timeline. It is futuristic and be able to take shocks; natural, circumstantial, global, regional, and domestic. The policy allocates resources so that the ends are achieved as per the stated objectives. The policy is a publicly declared statement by the leadership on which an open debate is undertaken. The policy depicts the mindset of the leadership and is reflective of peoples' aspirations. The policy must give a sense of security to the people, and must be progressive and forward-looking. Moreover, the policy must have the capacity to employ all available means (resources) to achieve ends identified by the leadership.

The term *strategy* originates from the Greek term *stratagem* meaning "the art of the generalship." The purpose of strategy is to optimally utilize the available resources, adopt the most preferred ways and achieve the assigned goals. Strategy is formulated by professionals after careful calculation of the probabilities of events that may be encountered toward the achievements of the goals set by the policy. Strategy springs into action once the policy has set the goals, allocated means and provided some broad guidelines on the available ways. Strategy must remain subservient to policy goals set forth by the political leadership. Strategy must remain confined to the available means and remain on course to achieve the set goals and must not creep its mission and lose focus. Strategy cannot afford to be bad or wrong because it would have deployed the available means. Strategy failure has a significant impact on the morale of the nation.

Whereas, the *Doctrine* is generally described as "ways of doing things" as per the principles developed through the knowledge of experience, and beliefs. It is developed after years of thought process and practical handling of doing the things in the right way. Doctrine addresses all the three components of war-fighting: material, moral and intellectual. Military Doctrine represents the collective advice on the best way of employing military forces in war. Doctrine contains a comprehensive account of the past practices, and the lessons learned.

Sun Tzu had prophesied on the need to have a well-thought-out strategy before launching any type or level of military campaign. However, he insists on the importance of the tactics as well. "Strategy without tactics is the slowest route to victory. Tactics without strategy is the noise before defeat."[1]

Who Decides What?

The next question would be who determines the security policy of a state. According to the Chinese sage Sun Tzu, it is the Emperor (the Chief Executive) who would identify the needs and allocate resources to achieve the assigned tasks. However, it would be the job of the Commander of the Armed Forces to employ the allocated means and make a suitable strategy to accomplish objectives.

Sun Tzu further illustrates his dicta that none would interfere in the job of the other, meaning that once the political leadership identifies the security needs and allocates the requisite means, then it is the responsibility of the military commander to ensure that the assigned tasks are accomplished to the best of their abilities. In case, the military commander thinks that the assigned tasks are beyond his material capacity, then he can always go back to the leadership for revision of the task or allocation of more resources.

It is understandable that material resources may not be adequate for the assigned tasks; therefore, it is incumbent upon the political leadership to assign only doable tasks and the requisite resources to demand the military's output. Concurrently, it is the responsibility of the military commander to ensure that his assessment about the strategic environment is based on sound and realistic information.

Therefore, it is essentially important that country's politico-military leadership correctly determines the strategic needs vis-à-vis available means. Any imbalance between the two will either compromise human security needs or the security needs of the state.

Going back into history, one notices that Pakistan's military leadership did not clearly outline the strategy adopted by its armed forces during the Kargil Conflict of 1999. In fact, the political leadership was on the path of reconciliation after the tit-for-tat nuclear tests in May 1998, followed by "Bus Diplomacy" by Prime Minister Vajpayee in February 1999, and policy of confrontational engagement did not exist at the time. Therefore, the military strategy also needed to be revised after due deliberations. However, the same was not done and military commanders planned their operations in isolation and without following the policy lines of the incumbent government.

Now, if some military officials claim that Kargil was a brilliant tactical plan because it achieved a perfect surprise with lot of territorial gains; they tend to ignore Sun Tzu's above precepts on strategy and tactics. In order to further elaborate this point, it is necessary to seek some guidance from another authority on strategy and tactics, Carl von Clausewitz. In one of his definitions, Clausewitz writes that strategy is the "[a]rt of using battles to win war," and in the same context, he defines tactics as the "[a]rt of using troops in the battle."[2]

Later on, President Musharraf, who was the military commander at the time of Kargil Conflict, dedicated one full chapter on the events in his book to outline his side of the story and clarifies a number of myths and realities. However, in doing so, at no stage did he make any mention of a strategy and insists on the brilliance of his tactical plans.[3] On the other hand, if the objective was to internationalize the Kashmir issue, it actually proved otherwise. According to the US diplomat Strobe Talbott, "Pakistan was almost universally seen to have precipitated the crisis, ruining the promising peace process that had begun in Lahore and inviting an Indian counteroffensive."[4]

Another aspect to which Sun Tzu insisted in this chapter was for the importance of clear direction to the military commanders under your command. However, in the case of Kargil, perhaps the same was the most ignored clause of Sun Tzu's dicta. For the sake of secrecy, the Kargil Plan remained with only few generals of Pakistan Army and not even shared with other components of Pakistan Armed Forces including PAF, which should have been an integral part of the joint planning, if the operation was inevitable.

It is necessary to reiterate that Sun Tzu said, "if state and armed forces are not on the same page, forget about war, if armed forces are not one, you cannot win war."[5] In modern-day wars, it is unthinkable that a war can be fought and won without integrating the air element at the planning stage. In fact, since the advent of air power in the early twentieth century, air power has been dominantly employed by all nations, be it the major wars like Gulf War-I (US-led allies against Iraq), or the Falklands War between the United Kingdom and Argentina in 1982. While the Gulf War-I was one-sided, but the Falkland War was fiercely contested in the air. "The war also illustrated the importance of air superiority—which the British had been unable to establish—and of advanced surveillance. Logistic support was vital as well, because the armed forces of both countries had operated at their maximum ranges."[6] Even though Argentinians could not sustain their initial gains, the British were stretched to limits, primarily due to logistics' tail that became too long to sustain the air battles.

Pakistan, too, could not have retained the captured posts without an active support by the air force, either in defense or as an offensive element. Perhaps, the right lessons were drawn and PAF was asked to lead the response to IAF strikes inside Pakistan near Balakot on the night of February 25–26, 2019. In fact, IAF strikes were in response to the Pulwama incident in which a Kashmiri youth blew himself near the convoy of Indian Security Forces, killing some forty-four personnel. India readily blamed Pakistan, like always, for planning and facilitating the attack, whereas Pakistan denied any involvement in the incident and officially condemned it as well. Responding to India's rhetoric of retaliating with punitive strikes deep inside Pakistan's territory, Pakistan's Prime Minister at the time, Imran Khan warned India of any misadventure. PM Khan in a televised statement stated that Pakistan would not think but react to India's action, if any.

IAF provided this opportunity to Pakistan too soon and within days carried out the strikes in general area Balakot, destroying a few trees at night, but claiming to have destroyed a terrorists' training camp killing over three hundred recruits. Pakistan refuted the claims and took foreign military officials and media personnel to the site that did not have any such facility. PAF responded to IAF strikes in line with the government's directives and destroyed two fighter aircraft; at least one of those fell inside Pakistan's part of Kashmir and the pilot Wing Commander Abhinandan was captured by the locals as he came down after a successful ejection from his burning aircraft. Besides these two kills by the PAF, IAF also lost a helicopter by its own fire in the fog of war in which at least six IAF personnel were killed. However, Pakistan did not claim shooting down the helicopter, though the same was shown burning by Indian media.

The results of two events, Kargil and Balakot, were diagrammatically opposite. Perhaps, it happened because Pakistan's politico-military leadership followed Sun Tzu's precepts on civil-military relations in latter's case, at least. Political leadership took the decision to respond and military led by the air element responded immediately to ensure that the adversary's plans of celebrating the victory on false claims are thwarted. Pakistan's response strategy proved successful and seeing a strong response from PAF, India decided not to escalate the crisis. Pakistan responded graciously and returned the captured pilot as a gesture of goodwill, thereby ending the crisis that could have escalated both horizontally and vertically between the nuclear neighbors.

This incident not only puts a seal on Sun Tzu's dicta about the supremacy of the political leadership on policy matters, but also highlights the significance of credibility of communication to the concept of deterrence. For the deterrence to be credible and effective, it is extremely important that the communication is clear, and capability is in place with unwavering intent. In this case, PM Khan's televised statement that we will not think but act to India's misadventure, was well supported by PAF with its prompt and measured response. Perhaps, the same response by PAF became instrumental in India's decision for restraint and avoidance of escalation, even though the events had the potential for horizontal as well as vertical escalation between the two nuclear neighbors.

Prepare Yourself, Do Not Wait for Enemy Attack

Utmost preparation for an impending war was another important aspect that Sun Tzu laid great emphasis on. Perhaps he was referring to another contemporary terminology: *deterrence*. The concept of deterrence is also as old as history of wars and conflicts. However, in modern times the concept, and definitions both, draw reference to the birth of nuclear weapons in 1945. Bernard Brodie was of the view that "if aggressor feared retaliation in kind, he would not attack."[7] He further explained: "Thus far the chief purpose of our military establishment has been to win wars. From now on its chief purpose must be to avert them."⁸ Other important definitions of deterrence included, "dissuading someone from an action by frightening that person with consequences of the action .... Deterrence depends what one can do, not or what one will do."⁹ However, in order to reach the level of capability and capacity to deter the adversary, it is extremely important to understand that it is earned after a concerted effort and through certain measures, and actions. Moreover, it is not a given fact and it is relative, and not across the board. In order to maintain its effectiveness, it must be sustained through continues research and development of technology, training, and strategy.

The former US Secretary of State, Henry Kissinger was of the view that,

> Deterrence is the attempt to keep an opponent from adopting a certain course of action by posing risks which will seem to him out of proportion to any gains to be achieved .... The higher the stakes, the more absolute must be the threat of destruction which faces him ... But, reverse is also true; the smaller the objective, the less should be the sanction.¹⁰

The concept of deterrence has many facets. "Deterrence can be a technique, a doctrine and a state of mind. In all cases it is about setting boundaries for actions and establishing the risks associated with the crossing of those boundaries."¹¹ Moreover, "... deterrence is mental. For deterrence to work you have to get inside your adversary's head."¹²

Deterrence is convincing adversaries that undesired actions would be responded in manners resulting in damages that may outweigh any likely benefit. The proliferation optimists, Kenneth Waltz leading, are of the view that the spread of nuclear weapons would deter the states from going to war against other states.¹³ "The weapons would, it is argued, provide weaker states with more security against attacks by stronger neighbors."¹⁴ However, this view is based on the assumption that states would think rationally and would not consider using these weapons, and therefore, nuclear arms races will not invite a nuclear exchange.¹⁵

Wilson is of the view that "[d]eterrence is limited in its application: it only works with those who stop and consider rationally the costs of what they're about to do. This means that from the outset, deterrence cannot be expected to cover all situations."¹⁶

Wilson further states: "Nuclear deterrence does not appear to be reliable or safe over the long run."¹⁷ Wilson's argument helped this research in ascertaining the efficacy of deterrence in a protracted conflict, particularly with rising asymmetries between India and Pakistan.

Where does Sun Tzu figure out in the debate of deterrence and the hybrid war? In fact, Sun Tzu's insistence on utmost preparations and winning the war without fighting signifies the concept of deterrence since the time immemorial. Moreover, his dicta on avoiding a war and forcing the decision without war refers to modern-day terminology of hybrid warfare. In case the two concepts are integrated wisely, winning the war without fighting might be accomplished with effortless ease. However, the modern-day global stakeholders are not willing to cede the concept of power for security under the paradigm of realism. Therefore, wars have been made to look unavoidable. Hence, it is important to understand the management of war to minimize damages to life and property, particularly of the non-combatants.

Conclusion

There is little doubt that in all civilized and democratic societies, all policy decisions on strategic matters come from the civilian leadership with allocation of resources, and then the strategy is formulated by the military commander to accomplish the defined ends. If the allocated means are not enough to accomplish the task or the assigned task is too steep and cannot be accomplished at this time, the Commander can ask the ruler to reconsider the two. A comparative analysis of the concepts prophesied by Sun Tzu and modern-day executions is placed in Appendix B.

Unfortunately, there is little effort on part of the international institutions and the global leaders to strive for peace and progress by avoiding wars and conflicts. The popularity that hybrid mannerism of modern wars is primarily to avoid international condemnation of the acts of violence in the absence of lawfare that could have posed certain restrictions on the perpetrators of the hybrid war.

Notes

1 Derek S. Reveron and James L. Cook quotes Sun Tzu in "From National to Theatre: Developing Strategy," *JFQ Issue* 70, 3rd quarter (2013): 113.
2 Paul Kennedy, ed., *Grand Strategies in War and Peace*, his article "Grand Strategy in War and Peace: Towards a Broader Definition" (Yale University Press, 1991), 1.
3 Pervaiz Musharraf, *In the Line of Fire: A Memoir* (Simon & Shuster, 2006), 87–98.

4 Adapted from Strobe Talbott's *Engaging India: Diplomacy, Democracy and the Bomb* (Brookings Institution Press). Talbott, former Deputy Secretary of State is the President of the Brookings Institution. Copyright © 2004, The Brookings Institution, accessed August 8, 2013, http://yaleglobal.yale.edu/content/day-nuclear-conflict-was-averted.
5 See James Clavell, ed., "Sun Tzu," in *The Art of War* (Dell Publishing, 1983).
6 Accessed January 24, 2023, https://www.britannica.com/event/Falkland-Islands-War.
7 Bernard Brodie, *The Absolute Weapon: Atomic Power and the World Order* (Institute of International, Studies, Yale University, 1946). Accessed November 15, 2013, http://www.airforcemag.com/MagazineArchive/Documents/2013/June%202013/0613keeper.pdf.
8 Bernard Brodie, "The Atomic Bomb and American Security," Yale Institute of International Studies, Occasional Paper no. 18, (Fall 1945). See also an expanded version of this paper in Brodie, *The Absolute Weapon*.
9 Kenneth N. Waltz, "Nuclear Myths and Political Realities," *American Political Science Review* 84, no. 3 (September 1990): 732–45.
10 Henry A. Kissinger, *Nuclear Weapons and Foreign Policy* (Harper & Brothers, 1957), 96.
11 Lawrence Freedman, *Deterrence* (Polity Press, 2004), 116.
12 Ward Wilson, "Deterrence in the 21st Century," November 20, 2013. Accessed September 8, 2014. http://www.publications.parliament.uk/pa/cm201314/cmselect/cmdfence/writev/deterrence/dic1.htm.
13 David J. Karl, "Proliferation Optimism and Pessimism Revisited," *Journal of Strategic Studies* 34, no. 4 (August 2011): 619–41.
14 See Edited extract from *Post-Cold War Conflict Deterrence* (Naval Studies Board, National Research Council, National Academy of Sciences, 1997). Accessed October 23, 2014, http://www.nap.edu/openbook.php?record_id=5464&page=R5.
15 Ibid.
16 Wilson, "Deterrence in the 21st Century."
17 Ibid.

· 5 ·

DEFEATING THE VICTORY

Introduction

This chapter deals with chapter IX and X of *The Art of War*. Sun Tzu insisted on the utmost preparation without which a defeat would be certain. His emphasis on the use of weather and terrain remains strategically important in contemporary era also. Sun Tzu also identified at least six plausible reasons for defeat: wrong assessment of the enemy, greed for power and command, poorly trained armed forces, unnecessary anger by the commanders, lack of discipline in the fighting force, and the poor usage of POWs.

Need for an Utmost Preparation

Sun Tzu insisted on the importance of utmost preparations for any military expedition which has become inevitable. "The general who wins the battle makes many calculations in his temple before the battle is fought. The general who loses makes but few calculations beforehand."[1] When Sun Tzu insists on "many calculations … before the battle," there is little doubt that he was emphasizing on thorough planning for an impending expedition, because he did not believe in waiting for the enemy to wrest the initiative.

In South Asian protracted conflict between archrivals: India and Pakistan, the Kargil planners may have given due consideration to Sun Tzu's dictates about planning to the last details, but the same could not be demonstrated when the situation began to unfold on the icy mountains. If one accepts Gohar's assertions that it was an old plan, then it was even more important to deliberate upon it more thoroughly in the changed paradigm. Several things had changed since India occupied the empty Siachen glacier in 1983, clearly in contravention to the Simla Agreement of 1972 between the two estranged neighbors. The two South Asian rivals had demonstrated their capability to develop nuclear weapons by carrying out the nuclear tests in May 1998, first by India and within days by Pakistan. Moreover, within one year of these tests, Indian Prime Minister Atal Bihari Vajpayee undertook the historic "Bus Diplomacy" trip to Pakistan and the two signed the *Lahore Declaration* which contained several Confidence Building Measures (CBMs). While the two governments were seriously revising their relationship for a peaceful and stable future, Pakistan's military leadership decided to go ahead with the Kargil Plan without reading the domestic or global environment.

For the elements of hybrid war to be effective against the target state, it is important that the campaign has an unrelenting support of the political leadership of launching country at the policy level. This is where Pakistan failed in the Kargil Conflict; however, such level of commitment by Indian political leadership was visible when it decided to support separatists' elements in erstwhile East Pakistan in 1971. India's deep involvement in the breakup of Pakistan through the hybrid deployment of integrated elements of hard and soft power was confessed by the incumbent Prime Minister Modi during his visit to Bangladesh in 2015. PM Modi takes pride in his arrest after participating in the Satyagraha movement at the age of 20–22. However, the concept of the Satyagraha movement as introduced by Gandhi was that of a non-violent movement against the evil, but Modi participated in a violent agitation movement which led to the breakup of Pakistan and created Bangladesh in December 1971.

In fact, Pakistan has been subjected to different forms of hybrid warfare over the last five decades, though the terminology gained ground only in the last two decades. Unfortunately, successive politico-military administrations in Pakistan failed to identify the levels of hybridity correctly, and therefore, the response prepared and executed was reactionary and based on evolving situations. Resultantly, Pakistan's stance was not recognized by the global community; instead, the perpetrators of hybrid war, led by India, were able to malign Pakistan for acts done by the elements against it. Therefore, it is necessary to correctly identify the levels of hybridity Pakistan has been subjected to, so that a coherent, sustainable, effective, and an integrated response is developed, and employed successfully.[2]

At the strategic level of hybridity, all elements of national security led by military establishments are deployed against the target state to achieve the stated objectives. After the creation of Bangladesh in 1971, to further impose existential threats to Pakistan, India carried out its nuclear tests in 1974. Concurrently, the US imposed sanctions on both India and Pakistan, which was far more harmful to Pakistan as it was refused an atomic reprocessing plant from France, sale of 110 A-7 aircraft from the US, and SAAB 37 Viggen aircraft from Sweden. In the early 1980s, India occupied the vacant and un-demarcated Siachen glacier, by employing the kinetic means of hybrid warfare.

The Siachen glacier is located in the Eastern Karakorum Mountain range, very close to the Chinese border in the south. It essentially forms part of Baltistan in the northern areas of Pakistan, and the same is reflected in major world maps.[3] The average height of the area is about 20,000 feet. Soon after their independence, Pakistan and India clashed militarily over Kashmir in what is usually referred as first India-Pakistan war of 1948. However, a Cease Fire Line (CFL) was established between the two as the outcome of UN-brokered negotiations. The Karachi Agreement was signed on July 27, 1949, and accordingly the Siachen glacier formed part of Pakistan's Northern Areas.[4] However, The Karachi Agreement did not consider formalizing the territorial boundary between India and Pakistan beyond the geographical reference of NJ9842, primarily due to inhospitability of the glacial terrain. Perhaps, the participants of the negotiations presumed that the area was not feasible for human living or any useful purpose.[5] The CFL established following The Karachi Agreement was renamed with minor modification as the LoC under the Simla Agreement of 1972. However, "The absence of a boundary line beyond the 'dead end' ... has led to the existence of a contested space ... no humans have ever lived, ..., or could live for protracted periods of time ..."[6] It is another battlefield where the Indian and Pakistani forces have been fighting a losing battle since 1984.[7] Siachen has turned into an ongoing conflict because Indian Army refuses to vacate the occupied land and controls it by maintaining a sizeable military presence. In fact, "territorial disputes occur when official representatives of one country make explicit statements claiming sovereignty over a specific piece of territory that is claimed or administered by another country."⁸ Saira Khan also quotes Staudenmaier that "new issues will arise during hostilities that plant the seeds for subsequent wars."⁹

At the operational level of hybridity, India deployed a combination of kinetic and non-kinetic elements of hybrid warfare in the post-nuclearized environment where an all-out conventional war became a distant probability. The same was exposed by European watchdog DisInfoLab through the India Chronicles, as explained in earlier chapters.

At the tactical level, India has been sponsoring the acts of terror in Karachi and Baluchistan for the last two decades, through the active support of local operators, and Indian infiltrators; also highlighted in the Indian Chronicles.

Having identified the levels of hybridity employed against Pakistan, not only by India but also by the US and its allies, through economic and military sanctions in international institutions, such as International Monetary Fund (IMF), and above all the FATF. Recent confession of India's External Affairs Minister S. Jaishankar clearly reflects that FATF was used as a political tool against Pakistan as part of a policy decision by India alongside its western allies. Therefore, to effectively counter known and unknown hybrid threats which would keep changing faces, Pakistan may formulate a multidimensional strategy which can counter the evolving threats at the corresponding levels of hybridity.

At the policy level, Pakistan may opt to respond to international organizations, institutions, and states collaborating against its vital national interests without any consideration of relationship, because "[a]ppeasement is a failed strategy and does not work in international affairs." To do so, Pakistan must take its allies and strategic partners in confidence to counter the concerted efforts of its enemies, be it in kinetic or the non-kinetic domain.

At the strategic level, Pakistan may respond by disrupting enemy's designs through the use of emerging technologies and expose the perpetrators with available evidence. Pakistan must not shy away from raising the levels even if global players are visibly supporting India in its efforts to malign it. Jaishankar's admission about FATF must be highlighted and the responsible institutions must be questioned about the processes adopted to keep Pakistan in the gray list for so long, despite its best efforts. Moreover, following Sun Tzu's dicta, Pakistan must not wait for India to act again, but remain prepared to counter any misadventure, as demonstrated in February 2019, following Balakot events.

At the operational level, apprehended must be summarily tried and punished to create deterrence in the hearts and minds of the potential terrorists.

India's apprehended spy, Jadhev, could be made an example because enough time has been given for his fair trials. Media campaign must be well orchestrated so that enemy's efforts do not succeed in demoralizing the people, and local facilitators in the media must not be allowed to sell enemy's agenda in the garb of freedom of expression.

At the tactical level, Intelligence-Based Operations (IBO) must be continued in areas identified as hotspots. Similarly, Counterinsurgency (CI), and Counter-terrorism (CT) strategies must be reviewed from time to time due to evolving threats and situations.

Khan outlined at least four aspects that were deployed against Pakistan to hurt its economy under the ambit of hybrid warfare: IMF, FATF, China-Pakistan Economic Corridor (CPEC), and Karachi. He was of the view that "the best strategy of the enemy is to erode the economic strength of the targeted country."[10] Khan asserted that Karachi, the economic hub and lifeline, was systematically destroyed to choke Pakistan's economy and make it dependent on international institutions. Pakistan has immensely suffered at the hands of International Financial Institutions (IFIs) by accepting unacceptable demands of raising the interest rates, devaluation of the currency, removal of subsidies, etc. At the moment, Pakistan is not only facing the risks of a possible default but also going through the highest inflation within the region as compared to Afghanistan, India and Bangladesh.

Pakistan is once again going through troubled waters due to evolving regional environment. India and its western allies are not interested in stability of Pakistan and indeed the region and therefore, are likely to continue their effort of hybrid warfare at all levels, warranting a matching response.

Need for Strategic Appraisal

Sun Tzu had insisted on the correct appraisal of the environment without which one's own security will be at a greater risk. "To secure ourselves against defeat lies in our own hands, but the opportunity of defeating the enemy is provided by the enemy himself."[11]

This phrase appears true in case of Kargil Conflict. Pakistan itself provided India a chance to claim victory, not only by using force to evict the intrusion, but also through a diplomatic offensive projecting Pakistan an aggressor and irresponsible nuclear state. Soon after the nuclear tests of 1998, Pakistan had the opportunity to reassert its stance on J&K and other disputes, as was recognized in the *Lahore Declaration* signed on February 21, 1999, during Prime Minister Vajpayee's visit to Pakistan. The *Lahore Declaration* outlined:

> Reiterating the determination of both countries to implementing the Simla Agreement in letter and spirit; Recalling their agreement of 23rd September, 1998, that an environment of peace and security is in the supreme national interest of both sides and that the resolution of all outstanding issues, including Jammu and Kashmir, is essential for this purpose.[12]

Whereas in India, its intelligence and army units were found napping and surprised, however, its political leadership took the lead to quickly put its house in order and tasked its military forces to get the Kargil Heights vacated, without actually going across the LoC. India did everything to push Pakistani soldiers back to their side of the LoC, but did not provide Pakistan any opportunity to play its nuclear card at any stage of the conflict.

Iraq's invasion of Kuwait on August 2, 1990, was even more disastrous when it comes to note for the correct appraisal of strategic environment. President Saddam Hussein's assumption that the world community would swallow his actions of invading a sovereign state (Kuwait) without a blink, proved fatal not only for the Iraqi nation but the entire region for decades to come. According to Dania Thafer, "From Kuwait's perspective, Iraq has always harbored an expansionist agenda towards Kuwait and their invasion fit into that agenda."[13] Particularly in the aftermath of its unnecessary war with Iran, after which Iraq was broke economically and considered Kuwait's oil as an easy target. Perhaps, Saddam misunderstood Sun Tzu's dictum of taking over enemy's land intact and that too without fighting, forgetting that he lived in different times and in a globalized world.

Iraq's invasion of Kuwait was a non-starter from the word go because it was at once declared as an end of Iraq by no one else but its own Army General Subhi Tawfiq, "Iraq's invasion of Kuwait was a dreadful day for both Gulf countries, but it was definitely the beginning of the end for Iraq."[14] Global community led by US under the umbrella of the UN swiftly moved to liberate Kuwait and the war which was started between UMPs (Iraq versus Kuwait), in fact ended between UMPs (Iraq versus the Rest). The ultimate loser was the people of Iraq who continue to suffer even after thirty years of the initial mistake by their long-time military ruler, Saddam Hussein.

Likewise, the outcome of Libyan conflict suggests that President Qaddafi failed to appreciate the gravity of the situation until it was too late. He not only lost his life but put his people at greater risk of disintegration of the state. However, in case of Yemen, it is evident that Saudi-led military alliance failed to evaluate the "capacity-to-resist" of the Houthis, and therefore got embroiled in an unending armed conflict, which has led to scarcity of food and medical supplies for the civilians. Sun Tzu focuses on the period before the war begins as a principle realm for a possible offensive strategy.

US-led allied invasion of Afghanistan is another eye-opener. Two decades of total control during which the entire physical and social infrastructure of Afghanistan was destroyed by the occupying forces, the final outcome remains, undesired. US-led allies exited Kabul in a haste following Doha Agreement signed on February 29, 2020, between the US and the Taliban, who were ousted after 9/11 for conspiring the attacks on the US. The Taliban government formed on August 15, 2021, remains unrecognized, and does what it wants against the world community's concerns about the treatment of women and girls' education.

The reason for the defeat of US-led allies in Afghanistan, as claimed by Taliban who did not give up the resistance against the occupation of their motherland, comes from the Sun Tzu's dicta: unnecessary anger. President Bush, without due consideration of time, preparation, and the outcome of any investigations into 9/11 attacks, ordered relentless bombings of Afghanistan followed by land invasion. However, the most advanced armed forces with most lethal weapon systems ever produced by the nations, could not tame Afghan resistance, and hence got defeated after spending over a Trillion dollars, losing thousands of soldiers, and two decades of its precious time, during which its newest rival; China has risen to a level that has become a cause of grave concern for the US, because it is now challenging the US for its sole leadership in global affairs.

Likewise, the poor usage of POWs cost the US dearly in terms of rights and morality. The US captured hundreds of Taliban fighters, but could not break their will to fight or tame them to defect the parent organization. Now, they are back in power in Afghanistan, boasting themselves as victors and doing as per their interpretation of the religion, politics, and state.

Conclusion

As the title of this chapter suggests, *Defeating the Victory*, fits well on Pakistan's intervention in Kargil in 1999, and the US invasion of Afghanistan from 2001–21. Lack of planning on part of the initiators, with unnecessary anger against the adversary, and wrong assessment of the opposing force's will to resist, created a compound effect and led to the defeat of a certain victory. In the case of Kargil, Pakistan failed to correctly assume the reaction of the Indian political leadership who asked its military unequivocally to get the intruders out. Pakistan did not anticipate the use of air power in the contest, perhaps because the same had been out of equation since the 1971 war. However, IAF played a decisive role in bombing out the intruding forces which did not have own air cover due to lack of synergetic planning.

In Afghanistan, unnecessary anger of the US leadership after 9/11, and wrong assessment of its military about peoples' resolve to resist the occupation, though the historical evidence existed, led to a humiliating exit of the most powerful military machine ever assembled against a UMP. The people of Afghanistan only repeated the history by displaying the same level of enthusiasm and resistance against the erstwhile Soviet Union, only two decades earlier.

Notes

1 Clavell, ed., "Sun Tzu," 11.
2 This author's "Identifying the Levels of Hybridity in War," *The Nation*, Islamabad, August 7, 2021.
3 Shireen M. Mazari, *The Kargil Conflict 1999: Separating Facts from Fiction* (Ferozsons, 2003), 3.
4 Ibid.
5 Faryal also has the similar opinion: "In 1949 the 'Karachi Agreement' and the 1972 'Simla Agreement' presumed that it was not feasible for human habitation to survive north of NJ9842. Prior to 1984 neither India nor Pakistan had any permanent presence in the area," accessed November 2, 2014, http://www.dawn.com/news/1141375/siachen-the-place-of-wild-roses.
6 Faryal Ali Gauhar, "Siachen: The Place of Wild Roses," accessed November 2, 2014, http://www.dawn.com/news/1141375/siachen-the-place-of-wild-roses.
7 Shamshad Ahmad, "Revising the Tide of History: Kashmir Policy—An Overview-II," *Dawn* (Karachi), August 6, 2004.
8 Sara Mitchell, *Territorial Disputes*, accessed 13 July 2021, https://www.oxfordbibliographies.com/view/document/obo-9780199743292/obo-9780199743292-0178.xml. The Issue Correlates of War (ICOW) Project has identified over 800 territorial disputes globally since 1816.
9 Saira Khan quotes William Staudenmaier from "Conflict Termination in the Nuclear Era," in *Conflict Termination and Military Strategy*, ed. Stephen Cimbala and Keith Dunn (West View Press, 1987), 15.

10 Ashfaq Hasan Khan, "Economic Coercion and Sabotage: A New Instrument of Hybrid War," in *Living Under the Hybrid War*, ed. Ashfaq Hasan Khan and Farah Naz (National University of Science & Technology (NUST), 2020).
11 Clavell, ed., "Sun Tzu," 19.
12 The Lahore Declaration was signed by the Prime Ministers of India and Pakistan on February 21, 1999, in Lahore.
13 Arwa Ibrahim, "Thirty Years on, Iraq's Invasion of Kuwait Still Haunts Region," August 1, 2020, accessed November 11, 2020, https://www.aljazeera.com/news/2020/8/1/thirty-years-on-iraqs-invasion-of-kuwait-still-haunts-region.
14 Ibid.

· 6 ·
SIGNIFICANCE OF INTELLIGENCE OPERATIONS

Introduction

This chapter analyzes the remaining chapters (XI, XII, and XIII) of *The Art of War*. These chapters reflect Sun Tzu's command on tactical considerations of topographical features for the offensive action. He also outlined the significance of intelligence operations that has revolutionized due technological developments and become an extremely useful element of hybrid warfare.

In fact, the rapid developments in applications of communication tools are making the human beings dependent on technology. The efficacy of these technological tools was tested to its limits during the Pandemic crisis, worldwide. Even the most advanced states with state-of-the-art medical facilities remained paralyzed for months, and had to resort to work from home; from executives to the clerical staff. The education institutions remained closed for nearly two years at some places and conducted online classes and Zoom webinars for an extended period, until the pharmaceutical industries in the developed world came up with the vaccines considered reliable to resume normal lives. At some places like Italy, Brazil and India, the mortality rate shot up during different phases of the Pandemic to the alarming levels and a total shutdown had to be observed for weeks. Likewise, manufacturing of small and large industrial goods remained under tremendous stress thereby affecting the supply chain management worldwide, causing demand and supply gap that negatively impacted inflation, and led to a price hike of essential food items.

Information Warfare

Information warfare is as old as the warfare itself. However, in the earlier times, it mostly relied upon misinformation, disinformation, psychological operations, propaganda campaigns, and deception. It had been successfully employed by all nations, perhaps since the days of Sun Tzu, who said, "All wars are based on deception."[1] The process of expanding and developing the informational type of warfare was concomitant to the technical evolution of communication. Hitler's rise and sophisticated propaganda, a subdivision of informational war,[2] was a clear demonstration of the same.

In modern times, Information Technology (IT) gained prominence with a study published by Harvard Business Review in 1958. The IT has revolutionized the communication, and enhanced interaction between people, groups, communities, and states. Since it was aimed at improving communication, and information dissemination, it expanded its area of coverage at a much faster rate than anticipated. Moreover, human appetite to absorb technology also helped spread its wings quickly, and across the globe in a short time, primarily because it helped bring people together, and facilitated them in education, health, travel, communication, and improved lifestyle.[3] Mitrovic also outlines at least two factors in this regard. "First, advances in technology, chiefly information communication technology (ICT), is an important driver that concerns digital capabilities and their associated security factors. Second, public attitudes have significantly changed towards military force, perhaps more in some states than others."[4]

Since the era of Sun Tzu, informational warfare had been an integral element of hybrid war that would precede any kinetic operations.

> The main overall objective of the informational war is to determine, to control or at least to alter the strategic decisions of foreign policy, security and defense, to pervert or hinder the instruments intended for the military component of a state, and to hinder the functioning, if not to block some elements related to the security of a state. The functioning of the informational war is done in an integrated manner, on all three dimensions, with deliberate dosages and a vast instrumentation built in time.[5]

No sooner the states started to improve their adaptability to newer technology, the relatively developed states started to weaponize them to gain advantage of the invention. Information warfare became exceedingly sophisticated and damaging. The phenomenal development in the domain of IT benefited electronic media the most, and perhaps the US-led Allied Forces successfully employed media as a tool to demonize Iraq, and its leadership to their advantage in the Gulf War-I. Likewise, India employed a successful media campaign against Pakistan in The Kargil Conflict of 1999.

A never-ending media war has since started by the relatively advanced countries led by the US. India too invested heavily on its human resource to make significant progress in the field of IT, and today, leads the domain in software development across the globe. Some of the global giants in IT are led by Non-Resident Indians, like IBM, Google, Microsoft, Adobe, etc. The human quest for better tools of communication continued, and led to the introduction of first smartphone by IBM, "the Simon Personal Communicator (SPC) in 1994."[6] Though it was not as sophisticated as today's smartphones are, but it opened the floodgates of newer technologies for Android and iPhones.

The phenomenon of social media appeared on the horizon in 1997, with digitization gaining significant ground in the ever-widening domain of IT. And, as smartphones became more accessible, the social media sites, and applications also made inroads in the society. As we stand today, the use of social media has become a household, and increasingly used to propagate one's own point of view.

Perhaps, the purpose of such rapid development of IT was to achieve better connectivity, integration, and networking in the globalized world. However, it was cleverly exploited by the relatively developed states for the purpose of warfare. The phenomenon is commonly referred as fifth generation warfare or the hybrid warfare.

Today, the social media stands fully weaponized, and is being aggressively employed against the target states. India is using it to project Pakistan as a facilitator, and a supporter of Taliban in their efforts to form a government in Afghanistan. India, in concert with certain elements in Afghanistan, and Pakistan, instigates, and propagates trends on social media, mainly on Twitter, to win support from its western allies against Pakistan. India created hundreds of fake websites, and accounts to project Pakistan as a state sponsoring terrorism, and engaged in money laundering which is then used for terror financing. Pakistan has been raising the issue of India's involvement in terror attacks, and submitted a dossier of evidence to the UN, but to no avail. However,

Pakistan's claim gained support when European Watchdog DisInfoLab revealed Indian Chronicles in 2020, highlighting India's deep involvement in running over 750 fake websites, and accounts to malign Pakistan in western capitals. Yet, India's propaganda campaign was successful in getting Pakistan's name in the gray list of FATF in 2017.

The social media, as a tool of hybrid warfare is successful, because of its accessibility, thanks to the development, and affordability of IT. Unfortunately, lack of awareness makes the people more susceptible to fall prey to false propaganda of the adversary. Moreover, these social media Applications are often engaged in sharing client data with state institutions. Recently, India's opposition parties accused Prime Minster Modi of compromising national security following revelations that dozens of Indians were potential targets of snooping by Israeli-made spyware. Sensitive information gathered through the social media is then manipulated by the state institutions to prepare, and launch malicious campaigns against the target state, and even interfere in the electoral processes.

Similarly, the Western world is using the social media against Russia in its war on Ukraine. The statements of the world leaders are making headlines to create strategic effects, while the images from the field are exploited at operational as well as tactical levels.

While the newer inventions, and developments in the domain of IT have brought people, societies, and states, together, but has put them at a greater risk of getting manipulated, and maligned. It is, therefore, necessary that social media be de-weaponized and held back from becoming a weapon of mass manipulation in the hands of the relatively developed states.

Farah Naz is of the view that in the evolved environment where social media plays a dominant role in shaping opinion, the most important thing is to construct a narrative that is appealing and attractive enough to bring change in people's thinking. She asserted the significance of a workable narrative that is proactive in its essence and based on sound footing aimed at unsettling the opponent's objectives. Naz insisted that "understanding narrative warfare is a necessary precondition for both comprehensive state policy and an informed public debate on issues, particularly security."[7]

Cyber Warfare

On the other hand, the cyber warfare in its present form may be a new phenomenon, however, in the olden days, the rivals made efforts to physically disrupt the communication systems of the adversary. Likewise, well-thought-out strategies were adopted to access the information systems and create mechanisms of misinformation in the enemy ranks utilizing all available means.[8]

While newly developed technological tools greatly helped human beings in their routine lives, it has concurrently put them at risks of being disrupted in terms of privacy and financial transactions. One such activity which is illegal, immoral, but widely practiced by criminally minded computer experts is referred as *Hacktivism*, a term used for a person who hacks a personal data of individuals with an intention to cause psychological, and financial harm. Likewise, the reports of cyberattacks on financial institutions syphoning off huge sums has become a routine. Although, no large-scale cyber-attack has been reported yet in the ongoing Russia-Ukraine war, but numerous systems related to economy, defense, business, and decision-making processes at all levels, remain vulnerable.

In contemporary environment, competition between interstate relations often unfold in physical, information, digital, cultural and cognitive arenas. However, direct military confrontations are not very common between the equal military powers. Alternatively, the tactics used by NSAs play a significant role in their political and strategic goals, and the efficiency of these tactics has often surpassed those of traditional military means.

Moreover, the concept of cyber threats has blurred the internal and external dimensions of state security and allowed less powerful state and non-state actors to amplify their attempts at influence. The changing nature of traditional concepts of armed conflict and war has also been identified as one of the factors that have encouraged the appearance of hybrid threats, which, as they lie outside the threshold of what is considered conventional armed conflict and war, are in a position to achieve their objectives by reducing military and civilian casualties.[9]

Today, the cyber warfare, due to its immense power of disruption in the communication systems, is considered more suitable tool of offensive action against its rival. This is extremely dangerous in the sense that such an action can paralyze the relevant system through disruption for the defined timeframe. These systems may include highly sensitive defense establishments including command and control systems, missile firing sites, air defense systems, and more seriously the decision-making mechanism at the strategic level.

Cyber war and cyber conflicts are becoming prominent types of hybrid threats in the changed paradigm.

> With the advent of digital technologies, the rate of cyber-attacks has increased, such that states now commonly employ cyber-attacks against their rivals. For example, in 2007, Russia used cyber-attacks to damage Estonia's internet infrastructure in retribution for its removal of a World War II Soviet war memorial from Tallinn.[10]

This attack certainly harmed Estonia's government in the domain of commercial and political systems, concurrently strengthening Russia's operations. Chifu is of the opinion that "… the evolution of modern Russian informational war includes elements of fake and forgeries, reflexive control and active measures."[11]

The employment of cyber warfare as part of a strategy to disrupt enemy lines of communication and cause harm to its potential capabilities has made state institutions highly vulnerable. Likewise, the personal security which is an essential element of human security also remains susceptible to disruptions in his/her communication, location, and financial transaction to mention the few. Therefore, it is incumbent upon international organizations to device legal procedures thereby ensuring that cyber warfare does not impinge upon an individual's rights and he/she is not deprived of his/her financial assets.

As cyberspace has become a novel arena for conflict, the Internet has become one of the most important fronts in hybrid warfare. As a military strategy theory, hybrid warfare was proposed by Frank Hoffman. "Sophisticated campaigns that combine low-level conventional and special operations; offensive cyber and space actions; and psychological operations that use social and traditional media to influence popular perception and international opinion."[12]

In fact, this theory blends irregular, conventional and cyber warfare and also leverages political warfare along with methods such as foreign electoral intervention, diplomacy, lawfare and misinformation. In contrast, cyber warfare concerns digital attacks deployed to attack nations, disrupt their digital networks and cause them harm.[13] Cyber warfare has been widely recognized as a novel and prominent type of hybrid threat that follows the fifth dimension of warfare, and many studies have described cyber warfare as concerted digital operations against IT infrastructure.[14]

Cyberspace is now one of the most preferred domains to conduct cyber warfare.[15] Moreover, as a hybrid warfare tactic, cyberattacks can directly affect civilian populations and demoralize the people with emotional, psychological, and economic fallouts, at times. Since the lawfare is either silent or extremely weak in its execution and processes, the state and non-state actors have greater margins for action against classic state powers. The manipulation through disinformation, misinformation, fake and fabricated news, hybrid attacks can considerably weaken peoples' trust in state institutions.

Cyber threats can be divided into four areas.[16] First, *cyber espionage* is deployed in political, economic and military spheres. Many states, notably China, Russia, Iran, and the US, routinely employ cyber-espionage tactics. States typically conduct cyber-espionage activities either directly using their intelligence services or through corporate agents. Second, *cybercrime* is usually conducted for quick profit, and according to The World Economic Forum, "illicit proceeds from criminal activity are estimated to account for 2–5% of global GDP, or $2 trillion."[17] This includes theft, fraud, and money laundering that are typically by terrorist organizations, organized crime and hackers. Third, *cyber terrorism* seeks to obtain a broad range of information from individuals. However, in this particular situation, state agencies are at the forefront alongside terror outfits. Finally, *hacktivism* targets digital services and can involve theft and unauthorized publication of information as well as terrorist activities. Hence,

> [t]he use of kinetic and cyber responses to international terrorism has increasingly challenged the traditional distinction between war and peace. This distinction replaced the concept of identifying new multimodal threats, many of which were not examples of interstate aggression in the past. These new threats to world peace and security pose a major threat to Western ways of life in the context of ongoing asymmetrical conflicts in Pakistan, Afghanistan, and Somalia. These new wars are like asymmetric conflicts: dichotomous choices amid resistance and ordinary war that challenge traditional concepts of war and peace.[18]

Conclusion

Hybrid warfare, due to its diversity in approaches and applications, poses serious threats to a stable international system, because usually the adversary is not visible and may employ NSAs to achieve its objectives against a relatively smaller and weaker states.

Therefore, the civilized world would be far more dangerous place to live without an effective framework against perpetrators of hybrid warfare. The effective structure which has legal, moral, and social framework, which are binding and without any discrimination for the size and strength of the state.

Social media, on the other hand, is being successfully deployed to market the aims and objectives of the NSAs. It not only projects the purpose of their struggle but also helps in extending invitation to join their organizations, particularly motivating the youth who are the optimum users of the media. Moreover, most world leaders now use social media to undertake strategic communication, as well as policy pronouncements. During the ongoing Russia-Ukraine war, even the operational commanders have used Twitter, YouTube, LinkedIn, etc., to update the recurrent situation.

Notes

1 See Griffith, *Sun Tzu's Art of War*.
2 Chifu and Anghel, "Hybrid Warfare," 32.
3 This author's article, "Social Media: A Weapon of Mass Manipulation," *Daily Times*, Pakistan, August 30, 2021.
4 Miroslav Mitrovic, "The Genesis of Propaganda as a Strategic Means of Hybrid Warfare Concept," *Vojno delo* 70, no. 1 (2018): 34–49.
5 Iulian Chifu, "The Pattern of Russia's Informational War 1," in *The Changing Face of Warfare in the 21st Century* (Routledge, 2017), 51.
6 Doug Aamoth, "First Smartphone Turns 20: Fun Facts about Simon," *Time*, August 18, 2014, accessed March 11, 2023, https://time.com/3137005/first-smartphone-ibm-simon/.
7 Farah Naz, "Narratives Warfare," in *Living Under Hybrid War*, ed. Ashafq Hasan Khan and Farah Naz (National University of Science and Technology (NUST), 2022).
8 This author's article, "Cyber Warfare: A Weapon of Mass Disruption," *Daily Times*, Pakistan, March 21, 2022.
9 Rauta, "Towards a Typology of Non-state Actors in 'Hybrid Warfare.'"
10 Angelica Rutherford, "The Interplay between Energy Security and Law and Policy on Green Energy Development: A Socio-legal Analysis," PhD diss., University of Liverpool, 2019.
11 Chifu and Anghel, "Hybrid Warfare," 13.
12 Eve Hunter with Piret Pernik, *The Challenges of Hybrid Warfare* (RKK, ICDS, April 2015) and *Military Balance 2015* (International Institute for Strategic Studies, 2015).
13 Cholpon Abdyraeva, *The Use of Cyberspace in the Context of Hybrid Warfare. Means, Challenges and Trends*, Working Paper (Österreichisches Institut für Internationale Politik, June 2020), 107.
14 James Cockayne, "The Fraying Shoestring: Rethinking Hybrid War Crimes Tribunals," *Fordham International Law Journal* 28, no. 3 (2004), Article 4.

15 Jan Almäng, "War, Vagueness and Hybrid War," *Defence Studies* 19, no. 2 (2019): 189–204.
16 Popescu, "Hybrid Tactics."
17 Stefan Ellerbeck, "Nearly Half of Businesses Are Being Hit by Economic Crime, with Cybercrime the Gravest Threat. What Can They Do about It?," *World Economic Forum*, July 26, 2022, accessed March 9, 2023, https://www.weforum.org/agenda/2022/07/fraud-cybercrime-financial-business/#:~:text=The%20World%20Economic%20Forum%20says,global%20GDP%2C%20or%20%242%20trillion.
18 Viorel Barbu and Cristian-Octavian Stanciu, "Hybrid Risks, Vulnerabilities and Threats against Romania," in *International Scientific Conference*, "Strategies XXI" (National Defence University, 2020), 17–27.

CONCLUSION

Let's admit that the social scientists are stuck with traditional theories of realism, liberalism, constructivism, deterrence, compellence, etc., and have miserably failed to bring out any notable theory that could have prevented conflicts and wars across the globe. Though the guidelines existed in the form of Sun Tzu's precepts of winning the war without fighting. While hybrid threats are replacing conventional threats in global security environments, still the kinetic applications by the powerful against the UMPs continues unabated.

On the other hand, the natural scientists and the technologists continue to surprise the mankind and have benefited more to the socioeconomic and sociocultural development of human security. Therefore, there is a dire need to inject fresh air of *Originality & Objectivity (O2)*, if the social scientists are sincere in contributing toward the academic literature. This proposal is not, by any means, to show disrespect for the teachers, scholars, and researchers, but only to remind them of their core responsibility. In my opinion, the primary objective of a social scientist's research should be to explore ways and means for the benefit of the society.

Hybrid warfare, as a concept and military strategy, is as old as the warfare itself. Though the term *hybrid warfare* was first used in 2002 by Major William J. Nemeth, when he examined Chechen Insurgency (Nemeth, 2002). It

CONCLUSION

employs a combination of political warfare alongside conventional warfare, irregular warfare and cyber warfare. It also deploys variety of methods to instigate the people of the target state through media campaigns propagating fake news, diplomatic maneuvers, manipulation of the workings of international institutions, and even interfering in its electoral system.

The Chinese sage, Sun Tzu also gets quoted extensively in the context of the hybrid mannerism of warfare in contemporary wars. Some 2,500 years ago, he had prophesied that the supreme acme of skill is to win the war without fighting. Breaking the will of the people of the target state would be the real victory instead of destroying them. Sun Tzu insisted on capturing the enemy forces intact so that those could be used later on.

The perpetrators of hybrid warfare aim to create synergetic effects on target states by utilizing a range of tangible and intangible avenues. While tangible elements of warfare include military hardware that is visible to an enemy, the intangible elements of warfare, through which an enemy is invisible and operates from within, are often far more damaging and effective. These elements comprise the hybrid warfare domain and have an exceptionally large canvass. For instance, economic warfare tactics may include disruption to stock exchanges and currency markets, both of which can destabilize and weaken a state.

It is necessary to understand that when the erstwhile Soviet Union disintegrated as a political entity, there was no dearth of security apparatus on its inventory, especially once the dissolution process started in 1988, and was completed within three years before it lost its identity in December 1991. Likewise, the illogical invasion of Kuwait in August 1990, resulted in Iraq's own destruction following a war between UMPs. Iraq was neither poor nor militarily weak that any of its neighbors could run over it.

Today, international confrontations unfold in physical, informational, digital, cultural, and cognitive arenas. Increasingly, the tactics used by non-military actors play a significant role in their political and strategic goals, and the efficiency of these tactics has often surpassed those of traditional military means. As a result, the term *hybrid warfare* has been proposed to encompass the new complex and spatial multidimensionality of these tactics. Moreover, within the framework of hybrid warfare, the open use of military forces often occurs only in the final stages of an attack, and then only under the jurisdiction of the existing international regulatory framework for peacekeeping and crisis management.[1] Moreover, the use of non-violent methods is the essence

CONCLUSION

of hybrid warfare and reflects the shift of modern conflict from military to non-military means.

Hybrid warfare, as a concept and military strategy has been effective since the advent of warfare. This essential objective of hybrid threats is to break the distinction between combatants and citizens while, at the same time, military objectives recede into the background. The tactics used in these types of conflict focus on disinformation and propaganda. They aim to exploit economic, political, technological and diplomatic vulnerabilities; break communities, national parties and electoral systems; and disrupt key infrastructure.[2]

Since military commanders are faced with a complex and multifaceted hybrid wars, it is necessary that they must "utilize a wide range of capabilities, including high-intensity conflict normal units, decentralized special forces, and sophisticated platforms for intelligence and technology operations."[3] Managing hybrid threats requires consolidated defense and security institutions with territorial control and ongoing coordination with global organizations and partner countries, primarily because "hybrid threats encompass cyber security, critical information systems, and critical services, undermining public trust in government institutions and deepening social divisions."[4]

Within two decades of the twenty-first century, there have been a number of wars between UMPs which were unnecessary and have virtually destroyed a number of relatively weaker countries with no plausible gains for the aggressors, except an increase in sales of the war machines produced by their MIC. The Russia-Ukraine war is not a war between two UMPs because Ukraine has the moral, military, and financial support of NATO as well as non-NATO allies. Therefore, I refer to this war as one between Near-Equal Military Powers (NEMPs). The war has now entered its second year with no end in sight, perhaps because the US-led NATO does not want an end to this war, and therefore, supporting Ukraine to stand firm and keep fighting for which the MIC is happy to provide unabated support. Apparently, the US objective is to keep Russia embroiled in an expensive war, similar to the Afghan War-I. Russia, on its part, is perhaps mismanaging the war by prolonging it, even though most of the ethnic-Russian bordering region is already annexed and administered autonomously.

Understanding the enemy's design and overcoming one's weaknesses is perhaps the most important aspect after the first bullet has been fired. Therefore, it is essentially important that country's politico-military leadership correctly determines the strategic needs vis-à-vis available means. Any imbalance between the two will either compromise human security needs or the security needs of the state. Therefore, in my view, an intense debate on the subject is essentially required, especially in a country like ours, where the security needs are also genuine and the economic outlook is also not very promising.

This century is only twenty-two years old, and we have already seen a number of major wars and limited conflicts. Unfortunately, the century began with the tragic events of 9/11, which were followed by an even more terrible overreaction that lasted for two decades. South Asian rivals, India and Pakistan, were at each other's throats in one of the longest and extremely tense face-offs. Iraq was destroyed for no reason except to complete the unfinished agenda of eliminating its leader Saddam Hussein and once it was done, the invading forces quietly left the country in rubble to be cleared by a hand-picked government.

Syria and Libya have been pushed back by decades, and perhaps would never be the same. Kashmir and Palestine are losing hope due to lackluster interest by international institutions and stakeholders in the resolution of disputes. The war in Yemen has however halted for the moment after eight years of meaningless war, which has brought hunger, poverty, famine, and chaos for the poor people of an already struggling state.

Perhaps, the evolving world order is already amid World War-2.5 because nearly half of the world's population is affected due to ongoing wars and conflicts around the globe, either directly or indirectly. Moreover, no nation at this point in time, no matter how militarily or economically strong, has a silver bullet to achieve a certain kill. Perhaps, this is one of the reasons why ongoing conflicts and wars are protracted and there is no outright victor at the end. For instance, in Afghanistan, the war lasted over two decades between the world's most advanced, professional, and best-equipped armed forces in the history of warfare and a disorganized non-state militia. The outcome is part of one of the darkest chapters of the US war history. Likewise, the war in Yemen between alliances of some GCC countries against a resistance group is an unnecessary war that has led to an extremely precarious humanitarian crisis in the region.

Therefore, global stakeholders must realize the gravity of the situation now, and craft an International Security Arrangement (ISA), which may lead to a moratorium on wars and conflicts between UMPs. I am not suggesting any kind of world government, but a break from unnecessary wars that only benefit the developed nations and ruin the developing states. Perhaps, a fifty-year break and an ISA between global stakeholders with guarantees by international organizations for strict compliance may then give us all time and resources to fight the real issues like Climate Change, Pandemic, recession, and other evolving threats including cyber warfare, terrorism, and extremism due to the rise of NSAs, and the rise of far-right sentiments in some European countries, and India.

It is necessary to insist on minimizing the damage and avoid prolonging wars due to the absence of a sound war management of which the exit strategy forms an important part. Perhaps, it is essentially required that social scientists lay greater emphasis on studying "War Management," so that the politico-military objectives are accomplished with minimum damage to the warring nations, particularly to the innocent non-combatants.

This is one reason why this book is dedicated to the "Soldiers and Civilians who lose their lives during Unnecessary Wars?"

Notes

1 Banasik, "Challenges and Threats for the International Security."
2 Kayhan, "How to Profile PYD/YPG as an Actor in the Syrian Civil War."
3 Raugh, "Is the Hybrid Threat a True Threat?," 1–13.
4 Barbu and Stanciu, "Hybrid Risks, Vulnerabilities and Threats against Romania," 17–27.

APPENDIX A

Term	Definition
Policy	*Policy* refers to the ends that conform to the purpose of a state. It is forward-looking and capable of absorbing natural, circumstantial, global, regional, and domestic shocks. Policies allocate resources so that ends are achieved as per stated objectives. A policy is a publicly declared statement by a leadership body for which open debate is undertaken. Policies aim to give a sense of security to citizenry and to be progressive and forward-looking. Finally, a policy must have the capacity to employ all available means (i.e., resources) to achieve its defined ends (Shamsi, 2020).

Term	Definition
Strategy	The purpose of *strategy* is to optimally utilize available resources, adopt the most preferred methods and achieve assigned goals. Strategy is formulated by professionals after careful calculation of the probabilities of events that may be encountered during the achievement of the goals set by a policy. Strategy commences after policy has established set goals, allocated means and provided guidelines on available methods. Strategies should remain subservient to policy goals set forth by political leadership (Shamsi, 2020).
Doctrine	A *doctrine* is described as "way of doing things" as per principles developed through knowledge, experience, and belief. Doctrines are developed after years of consideration and assessment regarding the appropriate way to conduct relationships. Doctrine addresses all three components of war: material, moral and intellectual. Doctrine sets out fundamental principles by which military forces guide their actions in support of national objectives. Basic doctrine describes how those principles could be employed by relating defense strategy to military activities. Doctrine is not immutable and it needs to be regularly reviewed in the light of development in history, theory and technology.
Conflict	*Conflict* is fighting between countries or groups of people. It can also be defined as serious disagreement and argument about something important.

APPENDIX A

Term	Definition
State	A *state* is a society that possesses coercive authority and is legally supreme over an individual or group that is part of that society (Laski, 1935). In a broad sense, the state is a set of political, administrative, and coercive institutions governed by an executive authority. It defines the political framework and territorial boundaries within which societies conceptions of themselves exist (Kohli, 1987). States emerge within societies, gradually taking on specific form and character.
Hybrid warfare	Hybrid warfare, as a concept and military strategy, is as old as the warfare itself. It employs a combination of political warfare alongside conventional warfare, irregular warfare and cyber warfare. It also deploys variety of methods, such as fake news, diplomacy, lawfare and foreign electoral intervention. The major elements of *hybrid warfare* (as opposed to conventional kinetic military methods of engagement) include irregular, information, cyber, economic, and psychological approaches. Each element of hybrid warfare contains the ability to strike across all domains and at all levels of warfare: policy, strategic, operational, and tactical.
Hybrid peace	*Hybrid peace* is the observance of mutual respect, productive engagement, accommodation of views, adjustment for religio-cultural leanings, and space for common interests, and urge for co-existence (Shamsi, 2021).

Term	Definition
Lawfare	*Lawfare* is use of legal tactics by a country against its adversaries, particularly by challenging the legality of military or foreign policy. While the term *lawfare* can have a double meaning, in either case it relates to the use of legal systems and institutions to achieve an end. Examples of lawfare include damaging or delegitimizing adversaries, wasting their time and resources, or winning public relations battles.
Cyber warfare	The aim of *cyber warfare* is to disrupt enemy communications and cause harm to their capabilities. Cyber warfare, due to its disruptive power is considered a valuable tool of offensive action. Affected systems may include sensitive defense establishments, such as command and control systems, missile-launch sites, air defense systems, and decision-making mechanisms at the strategic level. Shamsi referred to cyber warfare as "weapons of mass disruption."
Asymmetric response	An *asymmetric response* is one where an opponent reacts to your actions in a manner that is totally different to the one you are employing. The aims, people, equipment, and tactics he uses are so dissimilar to yours that they could well nullify your actions and require a change in your operations.
Campaign	A military *campaign* is a series of planned moves in a theater of operations that theoretically culminate in the defeat of the enemy forces in that theater.

APPENDIX A

Term	Definition
Combat plan	A plan issued for a combat operation that may be effective immediately for planning purposes or for specified preparatory action. It is not put into execution until directed by the commander in a separate order of execution or until certain specified conditions are determined to exist. When its execution is directed, a *combat plan* becomes, in effect, a combat order.
National Security	*National Security* is the condition of a nation, in terms of threats. It is also the confidence held by the people that the nation has capabilities and policy to pursue its national interest. It is a combination of national interests, national objectives, and national power.
Elements of National Security	It includes Military Security, Political Security, Economic Security, Environmental Security, Cyber Security, Energy & Natural Resources Security, and above all the Human Security.
Deterrence	*Deterrence* is the prevention of hostile acts by others through the threat of retaliation. Essentially the opponent must be made aware that any act on his part will cause a response that would hurt him and his interests greatly.
Diplomacy	*Diplomacy* is the process of managing relations between different nation-states. It is regarded as a professional skill.

Term	Definition
Disproportionate response	The aim of defense forces is to create the requirement for a *disproportionate response* from potential enemies. The idea is to structure your forces in such a way that an opponent would have to expend so many resources to achieve his aim that it would be very difficult if not impossible to mount a campaign against you.
Financial Action Task Force (FATF)	Established in 1989, to establish international standards, and to develop and promote policies, both at national and international levels, to combat money laundering and the financing of terrorism.
Information operations	*Information operations* are actions taken to defend and enhance one's own information systems and to affect an adversary's information and information systems. They use a range of aerospace platforms to gather the data required for the production of information on a particular target.
National aims & objectives	*National Aims* are conditions in future desired and aspired by a nation. *National Objectives* are broad goals based on principles designed to support, promote and protect national interests.
National interest	Any issue that has a potential to directly impact the pursuit of National Goals can be classified as *National Interest*. These can be under five major dimensions: Geopolitical, Military, Economy, Sociocultural, Science and Technology. Interests change little over time,

Term	Definition
	though they are modified by the domestic and international situation. They include sovereignty, the security of the nation, the preservation of institutions and culture, and the improvement of social and economic conditions. National Interests are classified according to their intensity or priority. *Vital or Survival* Vital national interests are conditions that are strictly necessary to safeguard and enhance states' survival and well-being in a free and secure nation. *Most Important or Very or Extremely Important Interests* Extremely important national interests are conditions that, if compromised, would severely prejudice the ability of the state to safeguard and enhance the well-being of its people. *Important Interests* Important national interests are conditions that, if compromised, would have major negative consequences for the ability of the government to safeguard and enhance the well-being of its people. *Peripheral, Secondary or Less Important Interests* Less important or secondary national interests are not unimportant. They are important and desirable conditions, but ones that have little direct impact on the ability of the government to safeguard and enhance the well-being of its people.

Term	Definition
National power	It is nation's ability to achieve its objectives and is defined in terms of elements of national power, which are mixtures of political, social, diplomatic, economic and military factors.
Nation	A large aggregate of people united by common descent, history, culture, or language, inhabiting a particular territory.
Nation-State	A form of political organization (State) in which a group of people who share the same history, traditions, or language live in a particular area under one government
Combat power	*Combat power* is the total means of destructive and disruptive force that an armed service can direct against an opponent at a given time. Combat power comprises land, sea and air power. It is a balance of the most appropriate forces required to undertake a campaign.
Combined operations	*Combined operations* are conducted by the armed forces of a number of nations, while joint operations are those undertaken by the various services of a single country.
Concurrent operations	*Concurrent operations* are a range of military actions that are undertaken simultaneously to achieve the maximum effect.
Communications security	The protection by all measures to deny unauthorized persons information of value that might be derived from a study or receipt of communications.
Command	An officer who commands an element of a military organization is responsible for controlling it, organizing it, and ensuring that it meets its commitments.

APPENDIX A

Term	Definition
Communications	*Communications* are the systems and processes that are used to send and receive information usually using radio and its associated systems. Communications are vital in ensuring that data and the analyzed information are received and disseminated in a timely manner.
Concentration of force	*Concentration of force* is the concept that maximum pressure will be applied to an enemy's weak spot, which is generally agreed to be his center of gravity.
Contingency	A *contingency* is something that might happen in the future, and it is a feature of military activities that plans are frequently devised to cover most possible contingencies. The word is sometimes used loosely to mean a crisis.
Deception	*Deception* is the deliberate act of making someone believe something that is actually false.
Decisive warfare	*Decisive warfare* is the use of all elements of combat power to defeat the enemy as swiftly as possible with overwhelming force.
Operations other than war (OOTW)	*Operations other than war* are those conducted in hazardous circumstances to relieve distress and improve security in a place where the local civil administration has broken down because of conflict or natural disaster. They include evacuation, diplomacy, humanitarian aid, and peacekeeping.

Critical definitions are tabulated here for the purpose of understanding on the terminologies frequently used in the book. Table is prepared by the author; however, the definitions are taken from published sources such as https://airpower.airforce.gov.au/; http://www.raaf.gov.au/; https://www.thepresidency.org/; http://homeport.seacadets.org/, and so on.

APPENDIX B

Comparative Analysis
Sun Tzu's Precepts versus Modern-Day Thinking

S. No.	Domain	Sun Tzu	Modern Day
1.	Political	Political objective is supreme	National interests are supreme
2.	Economic	Acquire the state without destroying it	Strangulate the target state through sanctions and IFIs
3.	Human	Do not kill but capture and utilize them on your side	Ruthless bombardment of cities and towns equally killing the non-combatants

S. No.	Domain	Sun Tzu	Modern Day
4.	Cultural	Understand the culture of the target state and respect	Invade the state's culture and destroy the societal fabric
5.	Informational	All wars are based on deception	Black out, Disruption, Disinformation, Fake News
6.	War	Avoid war as much as possible	War is an instrument of policy
6.	Military (Outcome)	Win without fighting	Win by destroying the adversary
7.	Military (Planning)	Utmost preparation through extreme secrecy	Plans are always ready and going into war is not considered as a dangerous act particularly against the UMPs
8.	Military (Duration)	No prolonged wars	Wars go on for years and years
9.	Military (Know the Enemy)	Thorough knowledge of enemy and own, through combat comparison	Modern technology is used to collect information on the enemy
10.	Civil-Military Relations	Political leadership gives the target and military leadership to accomplish it	Developed world is same but in developing countries, military has a bigger say

APPENDIX B

S. No.	Domain	Sun Tzu	Modern Day
11.	Environment	Weather and terrain were given due importance	Wars continue in all weathers and terrains due to technological revolutions
12.	Training	Lot of emphasis was given	Highly specialized now
13.	POWs	Treat humanely so that they can switch sides	UN Conventions but violations reported during and after all wars

BIBLIOGRAPHY

Conventions, Declarations & Reports

Aamoth, Doug, "First Smartphone Turns 20: Fun Facts about Simon." *TIME*, August 18, 2014.

Alexandre Alaphilippe, Gary Machado, Roman Adamczyk, and Antoine Grégoire. "Indian Chronicles: Deep Dive into a 15-Year-Old Operation Targeting the EU and the UN to Serve Indian Interests." *EU DisinfoLab*, December 9, 2020.

Darwich, May. "The Saudi Intervention in Yemen: Struggling for Status." *Insight Turkey* 20, no. 2 (Spring 2018).

The Lahore Declaration Was Signed by the Prime Ministers of India and Pakistan on February 21, 1999, in Lahore.

National Research Council. *Post-Cold War Conflict Deterrence* [Edited extract]. Naval Studies Board, National Academy of Sciences, 1997.

UN Human Rights Instruments. "Geneva Convention Relative to the Treatment of Prisoners of War," Adopted on August 12, 1949.

Williams, Nathan. "Ukraine War: Biden Says Nuclear Risk Highest since 1962 Cuban Missile Crisis of 1962." BBC, October 7, 2022.

Books & Journals

Abdyraeva, Cholpon. *The Use of Cyberspace in the Context of Hybrid Warfare. Means, Challenges and Trends*, 107 (Working Paper). Österreichisches Institut für Internationale Politik, 2020.

Ahmad, Shamshad. "Revising the Tide of History: Kashmir Policy—An Overview-II." *Dawn* (Karachi), August 6, 2004.

Al-Aridi, Alaa. "Legal Complexities of Hybrid Threats in the Arctic Region." *Teise/Law* 112 (2019): 107–23.

Almäng, Jan, "War, Vagueness and Hybrid War." *Defence Studies* 19, no. 2 (2019): 189–204.

Bachmann, Sascha Dov, and Andres B. Munoz Mosquera. "Lawfare and Hybrid Warfare—How Russia Is Using the Law as a Weapon." *Amicus Curiae* 102 (2015).

Bachmann, Sascha-Dominik Oliver Vladimir, and Hakan Gunnariusson. "Hybrid Wars: The 21st Century's New Threats to Global Peace and Security." *South African Journal of Military Studies* 43, no. 1 (2015): 77–98.

Barbu, Viorel, and Cristian-Octavian Stanciu. "Hybrid Risks, Vulnerabilities and Threats against Romania." In *International Scientific Conference, "Strategies XXI."* National Defence University, 2020.

Batyuk, Vladimir I. "The US Concept and Practice of Hybrid Warfare." *Strategic Analysis* 41, no. 5 (2017): 464–77.

Brodie, Bernard. "The Atomic Bomb and American Security." Yale Institute of International Studies, Occasional Paper no. 18 (Fall 1945).

Brodie, Bernard. *The Absolute Weapon: Atomic Power and the World Order*. Institute of International, Studies, Yale University, 1946.

Caliskan, Murat, and Michel Liégeois. *The Concept of "Hybrid Warfare" Undermines NATO's Strategic Thinking—Insights from Interviews with NATO Officials*, Small Wars & Insurgencies. Taylor & Francis, 2020.

Chifu, Iulian. "The Pattern of Russia's Informational War 1." In *The Changing Face of Warfare in the 21st Century*. Routledge, 2017.

Chifu, Iulian, and Gabriel Anghel. "Hybrid Warfare." In *The Changing Face of Warfare in the 21st Century*. Routledge, 2017.

Clavell, James, ed. "Sun Tzu." In *The Art of War*. Dell Publishing, 1983.

Cockayne, James. "The Fraying Shoestring: Rethinking Hybrid War Crimes Tribunals." *Fordham International Law Journal* 28, no. 3 (2004), Article 4.

Dvorak, Joseph. "Complexity in Modern War: Examining Hybrid War and Future U.S. Security Challenges." MSU graduate theses 3029, 2016.

Eisele, Ines. "Five Facts on Grain and the War in Ukraine." *Deutsche Welle*, January 11, 2022.

Ellerbeck, Stefan. "Nearly Half of Businesses Are Being Hit by Economic Crime, with Cybercrime the Gravest Threat. What Can They Do about It?." *World Economic Forum*, July 26, 2022. Accessed March 9, 2023, https://www.weforum.org/agenda/2022/07/fraud-cybercrime-financial-business/#:~:text=The%20World%20Economic%20Forum%20says,global%20GDP%2C%20or%20%242%20trillion.

Freedman, Lawrence. *Deterrence*. Polity Press, 2004.

Gauhar, Faryal Ali. "Siachen: The Place of Wild Roses." Accessed November 2, 2014, http://www.dawn.com/news/1141375/siachen-the-place-of-wild-roses.

Gray, Colin S. "The Changing Nature of Warfare?" *Naval War College Review* 49, no. 2 (1996), Article 3.

Griffith, S. B. *Sun Tzu's Art of War*. Oxford University Press, 1971.

Hoffman, Frank G. *Conflict in the 21st Century: The Rise of Hybrid Wars*. Potomac Institute for Policy Studies, 2007.

Hunter, Eve, and Piret Pernik. *The Challenges of Hybrid Warfare*. RKK, ICDS, 2015 and *Military Balance 2015*. International Institute for Strategic Studies.

BIBLIOGRAPHY

Ibrahim, Arwa. "Thirty Years on, Iraq's Invasion of Kuwait Still Haunts Region." *Al Jazeera*, August 1, 2020.

Isa, Mariam. "How Saudi Oil Attack May Impact SA." *Finweek*, September 26, 2019.

Jopling, Lord. "Countering Russia's Hybrid Threats: An Update." Draft Special Report, NATO Parliamentary Assembly, 2018.

Karl, David J. "Proliferation Optimism and Pessimism Revisited." *Journal of Strategic Studies* 34, no. 4 (August 2011): 619–41.

Kennedy, Paul, ed. "Grand Strategy in War and Peace: Towards a Broader Definition." *Grand Strategies in War and Peace*. Yale University Press, 1991.

Khan, Ashfaq Hasan. "Economic Coercion and Sabotage: A New Instrument of Hybrid War." In *Living Under the Hybrid War*, edited by Ashfaq Hasan Khan and Farah Naz. National University of Science & Technology (NUST), 2020.

Khan, Ashafq Hasan, and Farah Naz, ed. *Living Under Hybrid War*. National University of Science and Technology (NUST), 2022.

Khan, Saira. "Conflict Termination in the Nuclear Era." In *Conflict Termination and Military Strategy*, edited by Stephen Cimbala and Keith Dunn. West View Press, 1987.

Kikinezhdi, O. M., and I. M. Shulha. "Challenges of the Hybrid War: Gender in the Mass Media." International Scientific and Practical Conference, Vilnius, August 2019.

Kissinger, Henry A. *Nuclear Weapons and Foreign Policy*. Harper & Brothers, 1957.

Mazari, Shireen M. *The Kargil Conflict 1999: Separating Facts from Fiction*. Ferozsons, 2003.

Miroslaw, Banasik. "Challenges and Threats for the International Security as the Consequence of the Russian Federation's Hybrid War." *Science & Military Journal* 12, no. 1 (2017): 27–34.

Mitchell, Sara. *Territorial Disputes*. Accessed 13 July 2021, https://www.oxfordbibliographies.com/view/document/obo-9780199743292/obo-9780199743292-0178.xml.

Mitrovic, Miroslav, "The Genesis of Propaganda as a Strategic Means of Hybrid Warfare Concept." *Vojno delo* 70, no. 1 (2018): 34–49.

Mosquera, Munoz, B. Andres and Sascha Dov Bachmann. "Lawfare in Hybrid Wars: The 21st Century Warfare." *Journal of International Humanitarian Legal Studies* 7, no. 1 (2016): 63–87.

Mumford, Andrew. "The Role of Counter Terrorism in Hybrid Warfare." Report prepared for NATO Centre of Excellence Defence against Terrorism (CEO-DAT), November 2016.

Murat, Caliskan, and Paul Alexander Cramers. "What Do You Mean by 'Hybrid Warfare'? A Content Analysis on the Media Coverage of Hybrid Warfare Concept." *Horizon Insights* 4 (2018).

Musharraf, Pervez. *In the Line of Fire: A Memoir*. Simon & Shuster, 2006.

Naz, Farah. "Narratives Warfare." In *Living Under Hybrid War*, edited by Ashafq Hasan Khan and Farah Naz. National University of Science and Technology (NUST), 2022.

Nicu, Popescu. "Hybrid Tactics: Neither New nor Only Russian." *EUISS Issue Alert* 4 (2015).

Oleksandr, Moskalenko, and Volodymyr Streltsov. "Shaping a 'Hybrid' CFSP to Face 'Hybrid' Security Challenges." *European Foreign Affairs Review* 22, no. 4 (2017): 513–32.

Pusane, Özlem Kayhan, "How to Profile PYD/YPG as an Actor in the Syrian Civil War: Policy Implications for the Region and Beyond." In *Violent Non-State Actors and the Syrian Civil War*. Springer, 2018.

Raugh, David L. "Is the Hybrid Threat a True Threat?." *Journal of Strategic Security* 9, no. 2 (2016), Article 2.

Rauta, Vladimir. "Towards a Typology of Non-state Actors in 'Hybrid Warfare': Proxy, Auxiliary, Surrogate and Affiliated Forces." *Cambridge Review of International Affairs* 33, no. 6 (2019): 868–87.

Rutherford, Angelica. "The Interplay between Energy Security and Law and Policy on Green Energy Development: A Socio-legal Analysis." PhD diss., University of Liverpool, 2019.

Renz, Bettina. "Russia and 'Hybrid Warfare'." *Contemporary Politics* 22, no. 3 (2016): 283–300.

Reveron, Derek S., and James L. Cook. Quotes Sun Tzu in "From National to Theatre: Developing Strategy." *JFQ Issue* 70, 3rd quarter (2013).

Sadik, Dr. Giray. "Global Hybrid Threats and European Security in the Age of Trump, Growing Populism, and International Terrorism." *Europenow Daily*, November 2, 2018.

Shamsi, Zia Ul Haque. *Nuclear Deterrence and Conflict Management between India and Pakistan.* Peter Lang, 2020.

Shamsi, Zia Ul Haque. *South Asia Needs Hybrid Peace.* Peter Lang, 2021.

Stoker, Donald, and Craig Whiteside. "Blurred Lines: Gray-Zone Conflict and Hybrid War—Two Failures of American Strategic Thinking." *Naval War College Review* 73, no. 1 (2020), Article 4.

Talbott, Strobe. *Engaging India: Diplomacy, Democracy and the Bomb.* Brookings Institution Press, 2004.

Tanner, Jari. "Finland to Boost Security at Russia Border with Amended Law." *AP News*, July 7, 2022. Accessed March 14, 2023, https://apnews.com/article/nato-russia-ukraine-migration-sweden-moscow-002d36695a2e29f1d70d2c85faee225c.

Tóth, Gergely. "Legal Challenges in Hybrid Warfare Theory and Practice: Is There a Place." In *The Use of Force against Ukraine and International Law: Jus Ad Bellum, Jus In Bello, Jus Post Bellum*, edited by Sergey Sayapin and Evhen Tsybulenko, 18. Springer, 2018.

Trebilcock, Anne. "The ILO as an Actor in International Economic Law: Looking Back, Gazing Ahead." In *European Yearbook of International Economic Law 2019*, European Yearbook of International Economic Law, edited by Marc Bungenberg, Markus Krajewski, Christian J. Tams, Jörg Philipp Terhechte, and Andreas R. Ziegler. Springer, 2019.

Treverton, Gregory F., Andrew Thvedt, Alicia R. Chen, Kathy Lee, and Madeline McCue. *Addressing Hybrid Threats.* Swedish Defence University, 2018.

Vladimir, I. Batyuk. "The US Concept and Practice of Hybrid Warfare." *Strategic Analysis* 41, no. 5 (2017): 464–77.

Waltz, Kenneth N. "Nuclear Myths and Political Realities." *American Political Science Review* 84, no. 3 (September 1990): 731–45.

Wilson, Ward. "Deterrence in the 21st Century." *Crime and Justice in America 1975–2025* 42, no. 1 (November 20, 2013): 199–263.

Wintour, Patrick. "Iran and Saudi Arabia Agree to Restore Ties after China-Brokered Talks." *The Guardian*, March 10, 2023. Accessed March 14, 2023, https://www.theguardian.com/world/2023/mar/10/iran-saudi-arabia-agree-restore-ties-china-talks.

PETER LANG
PROMPT

Peter Lang Prompts offer our authors the opportunity to publish original research in small volumes that are shorter and more affordable than traditional academic monographs. With a faster production time, this concise model gives scholars the chance to publish time-sensitive research, open a forum for debate, and make an impact more quickly. Like all Peter Lang publications, Prompts are thoroughly peer reviewed and can even be included in series.

For further information, please contact:

editorial@peterlang.com

To order, please contact our Customer Service Department:

peterlang@presswarehouse.com (within the U.S.)
orders@peterlang.com (outside the U.S.)

Visit our website: www.peterlang.com

Prompts include:

Claudia Aburto Guzmán, *Poesía reciente de voces en diálogo con la ascendencia hispano-hablante en los Estados Unidos: Antología breve.* ISBN 978-1-4331-5207-8. 2020

William Robert Adamson, *Mine Own Familiar Friend: The Relationship between Gerard Hopkins and Robert Bridges.* ISBN 978-1-80079-485-6. 2021

Tywan Ajani, *Barriers to Rebuilding the African American Community: Understanding the Issues Facing Today's African Americans from a Social Work Perspective.* ISBN 978-1-4331-7681-4. 2020

Macarena Areco, *Bolaño Constelaciones: Literatura, sujetos, territorios.* ISBN 978-1-4331-7575-6. 2020

Robin Burgess (ed.), *FRANCESCO ALGAROTTI: AN ESSAY ON THE OPERA (Saggio sopra l'opera in musica) The editions of 1755 and 1763.* ISBN 978-1-80079-505-1. 2022

Desrine Bogle. *The Transatlantic Culture Trade: Caribbean Creole Proverbs from Africa, Europe, and the Caribbean.* ISBN 978-1-4331-5723-3. 2020

Jean-François Caron. *Irresponsible Citizenship: The Cultural Roots of the Crisis of Authority in Times of Pandemic.* ISBN 978-1-4331-8908-1. 2021

Jean-François Caron, *The Great Lockdown: Western Societies and the Fear of Death.* ISBN 978-1-4331-9535-8. 2022

Marcílio de Freitas and Marilene Corrêa da Silva Freitas, *The Future of Amazonia in Brazil: A Worldwide Tragedy.* ISBN 978-1-4331-7793-4. 2020

Mihai Dragnea. *Christian Identity Formation Across the Elbe in the Tenth and Eleventh Centuries. Christianity and Conversion in Scandinavia and the Baltic Region, c. 800–1600*, vol. 1. ISBN 978-1-4331-8431-4. 2021

Janet Farrell Leontiou, *The Doctor Still Knows Best: How Medical Culture Is Still Marked by Paternalism*. Health Communication, vol. 15. ISBN 978-1-4331-7322-6. 2020

Clare Gorman (ed.), *Miss-representation: Women, Literature, Sex and Culture*. ISBN 978-1-78874-586-4. 2020

Eva Marín Hlynsdóttir. *Gender in Organizations: The Icelandic Female Council Manager*. ISBN 978-1-4331-7729-3. 2020

Micol Kates, *Towards a Vegan-Based Ethic: Dismantling Neo-Colonial Hierarchy Through an Ethic of Lovingkindness*. ISBN 978-1-4331-7797-2. 2020

Sunho Kim, *Inner Mongolia, Outer Mongolia: The History of the Division of the "Descendants of Chinggis Khan" in the 20th Century*. ISBN 978-1-4331-8185-6. 2022

Feridoon Koohi-Kamali (ed.), *Exploring Roots of Inequality in Latin America and Peru*. ISBN 978-1-4331-8989-0. 2021

Guy Merchant, Cathy Burnett, Jeannie Bulman, and Emma Rogers. *Stacking Stories: Exploring the Hinterland of Education*. ISBN 978-1-80079-686-7. 2022

Matt Qvortrup, *Winners and Losers: Which Countries are Successful and Why?*. ISBN 978-1-80079-405-4. 2021

Peter Raina, *Doris Lessing – A Life Behind the Scenes: The Files of the British Intelligence Service MI5*. ISBN 978-1-80079-183-1. 2021

Peter Raina (trans.), *Heinrich von Kleist Poems*. ISBN 978-1-80079-043-8. 2020

Josiane Ranguin, *Mediating the Windrush Children: Caryl Phillips and Horace Ové*. ISBN 978-1-4331-7424-7. 2020

Dylan Scudder, *Coffee and Conflict in Colombia: Part of the Pentalemma Series on Managing Global Dilemmas*. ISBN 978-1-4331-7568-8. 2020

Dylan Scudder, *Conflict Minerals in the Democratic Republic of Congo: Part of the Pentalemma Series on Managing Global Dilemmas*. ISBN 978-1-4331-7561-9. 2020

Dylan Scudder, *Mining Conflict in the Philippines: Part of the Pentalemma Series on Managing Global Dilemmas*. ISBN 978-1-4331-7632-6. 2020

Dylan Scudder, *Multi-Hazard Disaster in Japan: Part of the Pentalemma Series on Managing Global Dilemmas*. ISBN 978-1-4331-7530-5. 2020

Wesley A. Stroud, *Education for Liberation, Education for Dignity: The Story of St. Monica's School of Basic Learning for Women*. ISBN 978-1-4331-7911-2. 2021

Geanneti Tavares Salomon, *Fashion and Irony in «Dom Casmurro»*. ISBN 978-1-78997-972-5. 2021

Zia Ul Haque Shamsi, *South Asia Needs Hybrid Peace*. ISBN 978-1-4331-9422-1. 2022

BIBLIOGRAPHY

Mohammad Rafiqul Islam Talukdar, *Local Government Budgetary Autonomy: Evidence from Bangladesh.* ISBN 978-1-80079-528-0. 2022

Shai Tubali, *Cosmos and Camus: Science Fiction Film and the Absurd.* ISBN 978-1-78997-664-9. 2020

Angela Williams, *Hip Hop Harem: Women, Rap and Representation in the Middle East.* ISBN 978-1-4331-7295-3. 2020

Ivan Zhavoronkov (trans.), *The Socio-Cultural and Philosophical Origins of Science* by Anatoly Nazirov. ISBN 978-1-4331-7228-1. 2020

www.ingramcontent.com/pod-product-compliance
Ingram Content Group UK Ltd.
Pitfield, Milton Keynes, MK11 3LW, UK
UKHW021249180426
11946UKWH00003B/41